Kister Distillation Symposium 2019 (Full Papers)

Topical Conference at the 2019 AIChE Spring Meeting and 15th Global Congress on Process Safety

New Orleans, Louisiana, USA
31 March - 4 April 2019

ISBN: 978-1-5108-8682-7

Printed from e-media with permission by:

Curran Associates, Inc.
57 Morehouse Lane
Red Hook, NY 12571

Some format issues inherent in the e-media version may also appear in this print version.

Printed by Curran Associates, Inc. (2019)

For permission requests, please contact AIChE
at the address below.

AIChE
120 Wall Street, FL 23
New York, NY 10005-4020

Phone: (800) 242-4363
Fax: (203) 775-5177

www.aiche.org

Additional copies of this publication are available from:

Curran Associates, Inc.
57 Morehouse Lane
Red Hook, NY 12571 USA
Phone: 845-758-0400
Fax: 845-758-2633
Email: curran@proceedings.com
Web: www.proceedings.com

TABLE OF CONTENTS

Francis Gregway Shinskey: 2019 and Beyond

Sigifredo Nino, P.Eng.

Summa Control Solutions Inc.

Montreal, QC, CANADA

SNino@SummaControlSolutions.com

Presented at the 2019 Distillation Symposium, AIChE Spring Meeting, New Orleans, Louisiana, March 31 – April 4, 2019

Francis Gregway Shinskey: 2019 and Beyond

Sigifredo Nino, P. Eng. (Snino@SummaControlSolutions.com)

Process Control Consultant, Summa Control Solutions Inc., Montreal, Canada

"Virtually any arrangement of loops will operate satisfactorily in the steady state - but is the unsteady state that tests control performance" (G. Shinskey 1967)

"There is no substitute for process knowledge, and certainly none for common sense" (G. Shinskey, 1994)

Abstract

The purpose of this paper is honour F. G. Shinskey, arguably the greatest contributor to Process Control for the Chemical Processing Industries and, in the process, to inspire the new upcoming generations of process control practitioners to keep his teachings close to their hearts. The paper will present several aspects of Shinskey's life, such as: His most important contributions, His awards, His family farm enterprise, His outspokenness with respect to the academic-industry gap, His musical skills and his paintings, and to narrate where his work and teaching brought him to; amongst other fascinating facts from his life that he was willing to share with this author.

In an age in which filling technical positions is becoming more difficult, engineering seems to have lost its luster (sadly). Interest in engineering from the new generations is waning and the few companies that are still looking for engineers are having a hard time finding them. Process Control expertise is also fading away. It is hoped that this presentation will inform new process control engineers to embrace Shinskey's contributions and instill in them an appreciation to work in the pursuit of a better world. Shinskey was always willing to share his findings and insights. He also encouraged academics and researchers to have a practical and approachable way to their inventions if they wanted to see them become part of the arsenal of tools available to process control engineers in the field. In this day and age, we have a proven engineering discipline; nowadays we need to leverage on the venues offered by the social media and the internet, in general, to carry on the message and timeless techniques Shinskey has gifted to us.

His essential bio

Shinskey was born in Tonawanda, New York on the 29th day of October 1931. He attended the Canisius High School in Buffalo, NY, where he graduated with honors at the age of 16. He graduated Magna Cum Laude as a Chemical Engineer from Notre Dame University (IN) in 1952. He served two tours of duty in Korea while on active service with the U.S. Navy and after working a few years for DuPont and Olin Chemicals, he went on to have a very prolific career at The Foxboro Company from 1960-1993.

While working at Olin Chemicals in Niagara Falls in 1958, he married his secretary Elizabeth Barret, who sadly passed away on December 2018. Eight children were born from his marriage. In Shinskey's own words: "the youngest is a Ph.D. specializing in infant cognition in London

(UK), the other ones live in WV, CA, FL, NJ and two of them in RI, where he as his wife have chosen to spend their last days."

His entrepreneurship

In an e-mail, exchanged on February 16, 2018 for the purposes of this article Shinskey wrote: "I restored a historic (1775) house in Foxborough, and operated the property as a family farm, with a large vegetable garden, raising a steer, pigs, chickens, turkeys, geese, and ducks to feed us year-round. That lasted only as long as we had a houseful to feed."

His artistic inclination

Throughout his lifetime, Greg also had many artistic pursuits. "I was a church choir director and tenor for many years and retired as a church organist at age 80. I began landscape painting with oils, continuing with watercolors and now working with pastels, of which you have some examples." Greg was generous enough to share some of his outstanding art with the author. The Figure 1 shows a couple of his painting, now owned by the author.

Figure 1 - A couple of Shinskey's Paintings

The globetrotter

He wrote a book out of his travel logbook: "TO THE ENDS OF THE EARTH: Travels of an Engineer", which he hoped to get published but it was rejected by the publishers; nevertheless, he wanted to share his experiences as a world itinerant; so herein I quote from the preface: "Though I have written many books, I am not a novelist or journalist, nor well known outside my technical field. By way of introduction, my specialty is Process Control. It is the technology applied to the automatic control of processes in power plants, chemical plants, petroleum refineries, paper mills, etc. I will not bore you with any of the technology; it simply happens to be the vehicle which brought me to the adventures described here. While I am not particularly well-known in my home town, my name is recognized in such faraway places as Australia, Singapore, Venezuela, Finland, and South Africa. In responding to invitations to lecture all over the United States, Canada, and over forty other nations, I lived these adventures."

Having become a lead authority, he was invited to lecture in several famous places such as the Royal Society, the Royal Institute in London, and Oxford University.

The awards

Shinskey's contribution to process control and engineering has been recognized several times, probably not at many as he should have had, probably due to his critical observations of the role of the academic institutions contribution to the preparation of engineers for the real plant challenges. Nonetheless, these are some of the awards he received:

ISA Applications Award (1977) for advances in pH control, ISA Education Award (1983), ISA Founder's Award (1988) for his time domain approach to process control, ISA Fellow (1990), ISA R.N. Pond Award (2003) for best paper of the year, ISA's Life Achievement Award (2008). Computing Practice Award from the AIChE (1992), Benjamin H. Bristol Fellowship, awarded by The Foxboro Company (1982), Sir Harold Hartley Medal, IMC, U.K. (1995), Nordic Process Control Award (1998), Control Engineering Practice Award, AACC (2000), Process Automation Hall of Fame (Control magazine, 2001).

In addition, he holds 17 U.S. Patents for inventions such as "Deadtime simulators for process control apparatus", "Fractionator control systems material balance computer and feedback control", and "Method an apparatus for characterizing and compensating for non-linear components".

The magician-engineer

According to Ed Bristol, Jim Douglas (U. of Massachusetts) spoke highly of Shinskey: "He walks into a room, puts on his black cape and pointed hat, waves his wand at the process and the damn thing works!". The author could not have put it better. His expertise and know-how are nothing short of magic.

Shinskey pioneered the control theory in the time-domain. This occurred after a request to give a course in process control to a heterogeneous group of professionals who would have difficulty following the Laplace domain and frequency domain theories. Those notes eventually became the first edition of his Process-Control Systems book published in 1967. Throughout the years, he expanded and improved this book into three additional editions, with the most recent version published in 1996, by McGraw-Hill in a "Chinese translation of internationally renowned textbooks in information technology and electrical engineering". His books were also translated to Spanish, Italian, Japanese, Romanian, and Russian.

In his time, Shinskey authored well over one hundred papers. His first: "For Gas-Phase Reactors… Design for Control of Temperature", was published in Chemical Engineering, October 5, 1959. And the latest: "Killing Model-Based Control Dead Time", published in Control Magazine in May 2013. If you tally that up, that is over half a century of invaluable contributions to the process control field in the form of articles that provide a wealth of information.

Figure 2 - PCS 4th. Edition in Chinese, 2014

In his time, Shinskey authored well over one hundred papers. His first: "For Gas-Phase Reactors... Design for Control of Temperature", was published in Chemical Engineering, October 5, 1959. And the latest: "Killing Model-Based Control Dead Time", published in Control Magazine in May 2013. If you tally that up, that is over half a century of invaluable contributions to the process control field in the form of articles that provide a wealth of information.

Some of Shinskey's contributions to process control (Not necessarily his inventions)

Shinskey's work in Distillation Control is probably one of his biggest contributions, having written two editions of his DC (1977 and 1982), which happens to perfectly align with this morning's session: Honoring the Contributions of Greg Shinskey: A Jubilee of Advances in Distillation Control.

"Critique of Chemical Process Control Theory" written by Alan Foss in 1973 where he stated that "the central issue to be resolved by the theories of process control is the determination of process control structure", probably, inspired Shinskey to research the subject. In 1983 he presented his Distillation Control methodology to illustrate his proposition on how to address control strategy design problem.

Based on that paper and the collaborations with Shinskey the methodology could be summarized as follows: First, understanding that a model is a simplified and reduced version of the reality, find a process mathematical model; it will open the possibility to better understanding the underlying behavior of the process and will allow for testing that may not even be possible on the actual process. Second, use Relative Gain Array as the principal means for pairing controlled and manipulated variables, supported by the simplification of the size of the multivariable problem provided by the dynamic response of the different process variables, e.g. flows and pressures have much faster dynamics than temperatures and compositions. Although relative gains (RG) can be estimated from plant testing, the availability of suitable differentiable models will allow the RG calculation using slopes instead of steady state gains, such as in

Shinskey's Distillation 4-components proposed method. Third, use partial decouplers, iff (If and only if) necessary. Fourth, apply feedforward to improve controls performance.

Shinskey's contribution spread wide and far among industries such as Heating Ventilation and Air Conditioning, Solar Energy Systems, Mineral Processing, Pulp and Paper, Fossil Power Generation, Oil and Gas, Oil refining, Petrochemical, and food industry. This demonstrates the universality and transversality of control theory and its application to chemical processes that Shinskey was able to exploit masterfully. There are many specific techniques that Shinskey either invented or developed, amongst them, Feedforward control as a high-performance controller (1963), Material Balance in Distillation Control (1963), Separation Factor in Distillation Control (1965), Inferential control (1968), External Reset (1971), Optimizing Control Performance (minimum Integrated Error) by minimizing P, I and Controller output (1972), Values of process control (1974), Development of the Bristol's RG (1967), Override controls (1967).

Shinskey on Energy Conservation – the entropy crusader

Let me begin by clarifying that entropy is NEITHER disorder NOR chaos (Lambert, 2002). Entropy is NOT a driving force, it is an index that measures the energy dispersion within the system and its surroundings; i.e. the distribution of energy among a large number of molecular motions relatable to quantized states. Shinskey has made clear that "energy conservation" is a misnomer, energy is always conserved, but that even if the energy flowing IN and OUT of a process remains equal, its QUALITY deteriorates; to the extent of becoming unusable.

Shinskey wrote (1978, pp.251)" Distillation is a classic example of a process using energy to create order. Typically, the products of a distillation column are in the same physical state as its feed and have the same energy content. Therefore, the energy use in the process simply passes through, increasing its entropy to reduce the of the products". Recalling a conversation where the author asked Shinskey if a highly entropic system would be more difficult to control, he replied "yes". No mathematical proof but an experience-driven statement. Although it was a loose comment, it has a lot of implications when it comes to the possibility of designing and building a process that might be uncontrollable. Process design and control should have the same standing during the design phases of a plant, something Shinskey was also an advocate of.

In the mid-'70s Shinskey became an "entropy crusader". From that period came his proposals, and recommendations, of how controls can be used to save energy and facilitate operation, amongst them, those for Distillation Control: Floating-Pressure Control, Heat-Pump Systems, Compressor-Expander Systems, Regenerative Heat Recovery (Feed Preheat with Bottoms product), Double-Effect Operation (Thermal Integration), and Coupling Unrelated Columns (Energy Integration).

The University-Industry gap

Shinskey was outspoken about the academic-industry gap throughout most of his career. Starting with his 1971 "To teach creativity" writing where he criticized the higher education institutions for their lack of contact with the real world and the consequent continuous proposition of solutions to problems that don't exist in the real life. In 1994, in "A critique of Education in Process Control" he decried the "tendency toward "dumb down" in all areas of education", and on his 2002 "Process Control: As Taught vs as Practiced" he questioned investigation solutions that "has led to research results left unused by industry and graduates left unprepared for industrial assignments", in the same paper he also criticized the study of control loops without nonminimum phase dynamics, overemphasis of setpoint response, the near exclusion of load regulation, and the omission of dynamics in the loop disturbance path.

Is the University-Industry gap closing?

On May 1st, 2018 the following message was posted in the AIChE Engage - Discussion Central:

"Chemical Engineering schools follow a fundamental knowledge approach in educating their students ... and I agree that this is the core mission of schools and I see the value in that ... The problem is the gap between the fundamental knowledge that fresh grads have and the operational knowledge needed by the industry ..."

I have seen strong resistance in academia towards incorporating some kind of operational knowledge/training on systems, equipment, and instrumentation that might accelerate the assimilation of young engineers into industrial operation environments post hiring ... I have been teaching chemical engineering topics at the "operational level" for the past two years and I can see the value to the industry in this approach ... The question is what does it take to convince "academia" to introduce a "healthy dose" of "operational training" that does not compromise the core mission of "fundamental knowledge-based education" for their engineering students?

Ammar Alkhawaldeh, PhD"

The post received 24 replies in a couple of weeks. So, the academy-operational knowledge divide is still an issue...

His legacy for the New Generations of Process Control Practitioners

Proven Process Control engineering is still needed today. It will also be needed in the future. For newcomers, it is quite relevant as they will be in charge of maintaining control strategies that he/she may have neither designed or have been a part of the implementation or commissioning team. In process control, the logic is backwards: Controls are guilty until proven otherwise; if a unit operation stops working all the sudden, it is the controls fault, even if they have consistently worked fine for years. The process control engineer has to prove that is not the controls by finding out what has really caused the problem. That being said, the engineer will need to be prepared to understand and troubleshoot both, the controls and the process. There is no free lunch on this front!

Education will be required in one way or another, as there will always be the need for in-house expertise in process control. Shinskey's books and papers will always be a source of valuable information whenever there is a need for an ingenious and practical solution to a process control problem. Do not get me wrong, theory is great. Theory allows for the exploration of alternatives not yet seen in the field. But theory cannot be the be-all-end-all to the success of a Process Control Engineer. It must be realistic, and applicable to real life. Keep that in mind.

Based on the commonalities between Shinskey's and my own experiences, as he pointed out on several occasions, the recommendation would be to follow his footsteps. He saw himself as a practitioner who managed to balance theory and practice. He gathered theoretical knowledge from academics like Edgar H. Bristol (1936-) and enriched his hands-on experience from practitioners like Carroll J. Ryskamp (1930-2016), whom he called "Mr. Distillation".

Shinskey always invoked Cyril Northcote Parkinson's law: "The time required to do a job will be equal the time allowed to do it" So make sure that you set reasonable limits to the delivery of your work, otherwise the work will never be finished, and the managers will not see the value of your contribution as process control engineer.

On Artificial Intelligence and other tools

The author wrote an expert system based on backward reasoning back in 1987. An expert in the field, a knowledge engineer and a computer programmer were required to build an expert system. To avoid all the risks associated to dividing the work among people from different disciplines, Shinskey himself wrote a Distillation Expert System back in the mid 80's. The author used Shinskey's tool: it provided help into getting to know what it takes to design distillation control strategies. Yet, there were cases where the software application could not be used as the column specifications would not fit the models and/or assumptions made in the expert system, it other words, even if it had the essence of Shinskey's expert knowledge of the subject it was not a substitution of him, by any stretch of the imagination. As a matter of fact, it was only when the fundamentals of distillation column design, operation and the understanding or underlying principles behind the expert system were grasped, that the use of the expert system became more "normal" to this author. Bottom line, a tool is not a substitution for either experience or knowledge.

In 2018 this author took an eleven weeks course on "Machine Learning and Data Mining"; software tools have remarkably improved over the past 30 years, however, the same "limitations" apply: "No magic (except for Shinskey's!), No free lunch". The data has to be either collected or extracted from a historian for what the purpose of the application is, the data has to be adequately selected, which can take 70% to 80% of the project time, the data mining algorithms must be understood, the resulting models need to be monitored and recalibrated, etc. All these is to give a clear message to the new generations of process control engineers: Tools are tools, and are not substitution for human expertise. Theory-Praxis Feedback in the learning process is priceless.

On persistence

The author's work with Shinskey for over 25 years, gave him the opportunity to learn from the man himself. At one facility, after years of the continued operation of a successful application of his material-balance controls on two heat-integrated distillation columns (Nino, 2011), the author was told by the primary contact at the facility, that nobody at that plant thought that it would work! No sure if he ever received that type of comment from any customer.

Shinskey firmly believes, as the author does, and as any process control engineer must know: There are lots of value and benefits to be accrued by applying sound process control strategies to the chemical processing plants. As far as Distillation Control is concerned: Shinskey told the author during the execution of a sidestream drawoff distillation column project: "Give material-balance controls a chance".

The moral of the story is: If you have a solid design, backed by strong engineering fundamentals, the trends on the DCS will speak on your behalf. All you need is the opportunity from the right person to prove the benefits or process control. Sadly enough, this is more often than not the hardest piece of the puzzle to solve. Finding a believer, in a position of authority is difficult. For this reason, persistence is a must-have in the Process Control Engineer toolbox.

References

Felton B. (2008) "Innovating, educating, writing, earn Life Achievement Award for Greg Shinskey", Intech Magazine, August.

Lambert F. L. (2002) "Disorder—A Cracked Crutch for Supporting Entropy Discussions", Journal of Chemical Education, Vol. 79 No. 2, February

McMillan, G. and S. Nino (2017) "The greatest source of process control knowledge", Control Magazine, October

Shinskey, F. G. (1967) *Process-Control Systems*, McGraw-Hill Inc., NY.

Shinskey, F. G. (1978) *Energy Conservation Through Control Systems*, Academic Press, New York (NY)

Shinskey F. G. (1979) "Entropic and Anti-Entropic Processes and the Future of Technology", Energy, Man and Civilization - Report on an Engineering Foundation Conference. Engineering Foundation, NY.

Shinskey, F. G. (1971) "To teach creativity", Instrumentation Technology, pp.34-38, October

Shinskey F. G. (1983) "From process knowledge to computer control: A methodology", Pacific Area Chem. Eng. Congress, Seoul, Korea, May 5-8

Shinskey F. G. (1994) "Process Control: Where have we been, where are we going", South Africa IMC symposium, Durban, June.

Shinskey F. G. (1994) "A Critique of Education in Process Control", IASTED Conference on Computers in Industry, Cairo, December

Shinskey F. G. (2002) "Process Control: As Taught vs as Practiced," Industrial & Engineering Chemistry Research, 2002, vol. 41, no. 16, pp. 3745-3750

DEEP UNDERSTANDING AND GREAT ACHEIVEMENTS OF PID CONTROL

Gregory K. McMillan, Emerson Automation Solutions, Chesterfield, Missouri

Biography

Greg retired from Solutia Inc. as a Senior Fellow after 33 years specializing in improving loop performance. Greg contracted as a consultant in DeltaV R&D for 12 years and is presently a Principal Senior Software Engineer in Simulation R&D for Emerson Automation Solutions focusing on dynamic modeling using the Digital Twin. Greg is the author of 80 articles and papers and 20 books. Greg's most recent books are *Advances in Reactor Measurement and Control, Tuning and Control Loop Performance 4th Edition, Good Tuning: A Pocket Guide 4th Edition*, and *Process/Industrial Instruments and Controls Handbook 6th Edition*. Greg is an ISA Fellow and received the ISA "Kermit Fischer Environmental" Award for pH control in 1991, the Control Magazine "Engineer of the Year" Award for the Process Industry in 1994 and was inducted with Greg Shinskey into the "Process Automation Hall of Fame" in 2001. Greg was honored as one of InTech's 50 most influential innovators for advancing automation and control technologies and received the ISA Life Time Achievement Award in 2010.

Abstract

Shinskey's articles and books are the best source of what is truly important and achievable by PID control. His extensive analysis of process relationships and dynamics and PID controller capability are unmatched and is the key to process control systems that can increase process capacity and efficiency. He has been the biggest influence in my career. Many of accomplishments can be traced back to knowledge gained from Shinskey.

Process metrics such as integrated error and peak error can readily be estimated for disturbances from the controller tuning settings showing the impact of poor tuning. The disturbances are not on the process output as what is commonly shown in publications but are on the process input. The implications in terms of focusing on what is truly important and eliminating distractions from special algorithms supposedly better than PID depend upon this fundamental understanding. PID feedback and feedforward tuning settings and tests should be based on recognition of the location and dynamics of the disturbance. Any desired setpoint response can be achieved by the simple addition of a setpoint lead-lag. The use of external-reset feedback can automatically prevent problems with cascade control, poor control valve response and the use of analyzers with variable sample or cycle times. Furthermore, dead time compensation can be more effectively achieved and its limits recognized. Simulation can be used to develop the best automation system and control strategies. The benefits gained from the insights should continually grow. It is our job to make sure this is the case.

Introduction

There is no way to convey the incredible body of knowledge developed and documented by Shinskey. Nearly all of the knowledge stems from a deep understanding of the process, dynamics, PID capability, and the role of tuning. His publications address and eliminate the many misconceptions that unfortunately still prevail. If there was just one wish to be granted by some magical power, it would be that every academic and practitioner would know and use what Shinskey knew.

Sample of Many Great Achievements by Shinskey

The sample of key learning points offered are indicative of fundamental principles that are essential to improving the performance of control loops for nearly all types of processes. Understanding these points can get us all going in the right direction. This can lead to much greater recognition of the value of control and optimization in improving process performance.

Power of external-reset feedback to reduce oscillations

By simply turning on the PID option for external-reset feedback (e.g., dynamic reset limit) and sending back to the PID a fast update of the actual process variable, valve position or speed being manipulated, the PID output from integral action will not change faster than the secondary loop, valve or variable frequency drive (VFD) can respond.

This capability reduces oscillations from valve backlash and stiction, poor positioner sensitivity, slow actuators, VFD setup deadband and rate limiting, violation of cascade rule (primary loop 5 times faster than secondary loop) and interaction. It can improve batch control and override control. It also provides directional move suppression by the simple addition of PID or analog output setpoint up and down rate limits to prevent unnecessary crossings of the split range point, fast opening and slow closing surge valves, less disruption from high PID gain action, and smooth gradual optimization by valve position control with a fast getaway for abnormal conditions.

Dead time compensation that is much simpler

By simply inserting a dead time block in the external-reset feedback path back to the PID, you can achieve dead time compensation as effective as a Smith Predictor but without the need to insert an open loop process gain or time constant into a model. Also, there is no loss of the process variable in the operator interface and the dead time can be updated as function of process operating conditions such as production rate. Shinskey in his tests eliminated many widespread misconceptions such as the benefit being greatest for dead time dominant processes (greatest benefit is for lag dominant processes), the PID must be more aggressively tuned to see the benefit, the compensator can do better than MPC, and the dead time compensator is more sensitive to a block dead time larger than actual dead time. In a regular PID controller, larger than actual dead time used in tuning does not create oscillations.

Integrated error metric as a function of tuning, scan time and filter time

Missing in nearly all of the literature is the key fact that the integrated error for a load disturbance is a function of tuning settings. Studies showing the benefits of algorithms developed to replace the PID typically fail to take this into account allowing the developer of the algorithm to tailor test results to show the supposed benefits of the new algorithm. The equation developed by Shinskey is incredibly elegant. Shinskey extended it to include the effect of scan time and filter time by simply adding them to the reset time in the numerator.

Realization that most disturbances affect a process input

Perhaps the greatest disconnect between academia and industry stems from not realizing that most disturbances are on the process input instead of the process output. The consequences are greatly increased by studies that use balanced self-regulating processes where the primary process time constant is about the same size as the process dead time. The result has been algorithms and tuning methods that would be detrimental to the point of being hazardous for many of the more important composition and temperature control loops in the chemical industry. These loops tend to have a near-integrator or true integrating process response and in some cases, a runaway response demanding a much more aggressive algorithm and tuning. For example, some experts whose world was only balanced self-regulating loops advocated tuning to prevent overshoot by the controller output of its final resting value. This would be unsafe for near-integrating, true integrating and especially runaway processes. Shinskey has written about open loop unstable processes (e.g., parallel heat exchangers, steam jets, and highly exothermic batch polymerization reactors).

Tuning first for good load response and then good setpoint response

Nearly all tuning methods tested by making a setpoint change when in fact the tuning objective should in most cases be tuning for dealing with unmeasured load disturbances. The test for such disturbance is quite straightforward and simply involves momentarily putting the controller in manual, making an output change and putting the controller back in auto. The tuning should minimize integrated and peak error with minimal oscillation. Good setpoint response can be obtained by inserting a lead-lag on the setpoint. The lag is simply set equal to the reset time and lead is set equal about ¼ the lag time. Using a lead helps get the process variable to setpoint faster. If this is not a requirement (e.g., bioreactor pH shifts), the lead time can be left at zero, which is equivalent to a PID structure of PD on PV and I on E.

Controlled and manipulated variable pairing

The best pairing is normally the controlled and manipulated variables with a relative gain array value near unity. For integrating processes, the process variables are converted to a variable with a steady state being a constant ramp rate by simply using the rate of change of the process variable. Digital Twin simulations are good at finding the relative gain array. Testing in the field should be done to confirm choices. Sigifredo Nino, Shinskey's protégé is excellent in using RGA particularly for two point distillation column control. Shinskey developed many other criteria based on first principles, such as reactor endpoint control. A reactor pressure controller manipulating a gaseous reactant flow for a liquid co-reactant and liquid product and a reactor level controller manipulating a liquid reactant flow for a gas co-reactant and gas product will inherently manipulate the reactant in the right stoichiometry.

Dead time from discrete updates

There are a multitude of studies on the effect of I/O scan time, PID execution rate, wireless update rate and analyzer cycle time. The results are usually not truly representative and cannot be extended to other cases. The analysis could be simplified to the inherent effect simply being the ½ the time between updates plus latency. For scan times and wireless update rates, the latency is zero. Thus, the dead time is simply 1/2 the scan time or update rate. For PID execution rates, the latency can be significant if there is a lot of computation going on between the analog input and output besides the PID. For analyzers where the result is computed at the end of the cycle, the latency is the cycle time. The resulting dead time is then 1½ times the cycle time. Of course, there is also the sample transportation delay.

Simulation can quantify PID and dead time compensator performance

Shinskey extensively used simulation to show PID performance and robustness was affected by tuning and dead time compensation for various types of dynamics. This experimentation helped dispel many myths and verify the value of the PID and inherent tradeoff between performance and robustness. this concept is tended by modeling on a first principle basis challenging processes including automation system dynamics (e.g., delays, lags, deadband, resolution, sensitivity ...). For pH, Shinskey's charge balance is extended to include conjugate salts to detail the equipment and piping besides PID design requirements. Most recently, the Digital Twin is used to eliminate the errors associated with emulation and make the exchange of control system configuration and operator graphics seamless.

Block Diagram Showing External-Reset Feedback

Pneumatic controllers had to implement integral action by the use of a bellows and restrictor creating a filter whose time constant was the reset time in what is a positive feedback implementation of integral action. When electronic and digital controllers were developed most suppliers decided integral action could now be done with an integrator as seen in textbooks. Shinskey knew better and made sure Foxboro kept the use of a filter with a positive feedback implementation. Since most systems used a Series PID Form, his books showed this implementation. The time domain block diagram shows this capability along with eight different structures in a Parallel PID Form in DeltaV given as the default "Standard Form". A key to understanding this Form is that that when the error is zero, the input and output of the filter are equal suspending positive feedback. Also, integral action cannot be faster than the actual response of manipulated loop, valve or VFD in the external-reset feedback of the manipulated variable (MV) whether a process variable, stroke or speed. If the MV is not responding, integral action is suspended. Showing how to implement the parallel PID for a setpoint weight factor (β) requires numerically backing it out in the contribution to the output.

Examples Where External-Reset Feedback Helps

I find more and more situations where external-reset feedback suppresses oscillations or enables a more intelligent PID action. The fundamental capability by simply turning on this option is largely overlooked as seen in the fact that the PID algorithms in most DCSs and PLCs do not have the positive feedback implementation of integral action that gives external-reset feedback as an inherent capability.

Slow secondary loop in cascade control

While it is certainly better to design a cascade control loop so that the secondary loop is at least five times faster than the primary loop, turning on external reset feedback can reduce oscillations as seen on slide 5 for a setpoint change to the primary loop. Undershoot is reduced from 5.2% to 1.5% and the settling time is reduced from 1200 to 275 seconds.

Limit Cycle from backlash (deadband) in system with two or more integrators

A limit cycle will develop from backlash (deadband) if there are two or more integrators in the automation system and the process whether due to cascade control, a positioner with integral action, or an integrating process (e.g., gas pressure, batch composition or temperature, or level). Deadband can also exist from configuration in split range blocks and in VFD setup to reduce the response to noise. Turning on external reset feedback can slow down and the reduce amplitude of the oscillations in the process variable as seen in slide 6 for level.

Limit Cycle from stiction (resolution) in system with two or more integrators

A limit cycle will develop from stiction (resolution) if there are one or more integrators in the automation system and the process. There is almost always at least one integrator and the use of valves and digital systems ensure there is a resolution limit. Consequently, there is always a limit cycle albeit quite small for good throttling valves and a modern DCS or PLC with properly spanned transmitters. The standard I/O cards for a VFD speed setpoint and older DCS thermocouple and RTD I/O cards can have poor resolution creating a noticeable limit cycle. Turning on external-reset feedback can stop the limit cycle for a single point of integration but with an offset as seen in slide 7 for a flow loop.

Slow valve slewing rate or VFD speed rate of change limiting

Large actuators without a volume booster on positioner output and VFDs with a speed rate of change limit introduced due to an over concern of motor overload create a problem not seen for small changes in PID output. There is a burst of oscillations for large changes. Turning on external-reset feedback suppresses the oscillations as seen in slide 8 for a valve.

Poor positioner sensitivity (response time increases by order of magnitude)

Most people are unaware their positioner has poor sensitivity because tests normally use changes in signal of 1% or more. If the signal change is less than 0.5%, many positioners develop a response time that increases by a factor of 10 or more (e.g., 40 instead of 2 seconds). The response time also tends to double for a reversal of direction (e.g., 80 seconds). The solution is to turn on external-reset feedback of a fast readback of actual valve position and increase the PID gain to make larger changes in the PID output seen in readback.

Fast opening and slow closing surge valves

The addition of up and down setpoint rate limits to the analog output block can enable a fast opening of a surge valve to prevent surge and a slow closing to ensure a more gradual approach to more optimum operating conditions. In the past this was done in the field, with limited adjustability and visibility. The controller has to be tuned for the slowest response. By using external-reset feedback, tuning for fastest response can be used and the setpoint rate limits can be visible on the operator interface and accurately set in the PID configuration.

Unnecessary split range crossings

The discontinuity and nonlinearity at the split range point tends to make loops oscillate at this point resulting in detrimental consequences in terms of variability, loss of efficiency and in the case of bioreactor pH control, high osmotic pressure causing cell death from an accumulation of sodium ions form the unnecessary addition of sodium bicarbonate. Patience can be achieved by adding a setpoint rate limit slowing down a change in the use of a utility or reagent in the direction of with the least safety and process performance risk. External-reset feedback of the actual manipulated variable prevents the PID form continuing to drive the setpoint in the less desirable direction. This extra patience reduces reaction to short term situations and disruptions at the split range point.

Slow approach to optimum and fast getaway in valve position control

The addition of setpoint up and down rate limits provides adjustable and visible directional move suppression. For valve position control that maximizes feed rate or minimizes utility use by pushing user valves to a maximum or minimum throttle position, move suppression in the direction of increasing feed rate or decreasing utility use is done smoothly and gradually. However, if there is load upset or some undesirable situation, the move suppression in the opposite direction is made much faster to provide a fast getaway. External-reset feedback enables directional move suppression without the need to retune the PID.

Upset to other loops due to aggressive gain setting

In many applications the more immediate response of proportional action is needed to deal with load disturbances for process safety and performance. This is especially the case for near-integrating, integrating and runaway processes (e.g., composition, gas pressure, and temperature control). However, rapid movements of the manipulated variable are upsetting to the operator and can be upsetting to other loops particularly those sharing the same header or resource. The addition of setpoint rate limits to the analog output block or secondary loop with external-reset feedback of the actual manipulated stroke, speed or process variable can be optimized to meet loop objectives without upsetting other loops or the need to retune the PID.

Preload for best batch operation setpoint response

The time to reach setpoint without overshoot can be optimized by a smart preload response that puts the PID output at intelligent positions during the setpoint response providing a smooth response from P or PD to PID control.

Smooth and intelligent transition to selection in override control

The filter in external-reset feedback smooths out disruptions from noise, discontinuities and nonlinearities in override control. The unselected PID output integral mode contribution gradually moves to match the manipulated variable. This provides an ability to understand and optimize the contribution of proportional mode to provide the best response for selection.

Load Performance Visual Actualities

Shinskey showed that the optimum response to a load disturbance is simply a triangle where gains smaller than ultimate limit cause a higher triangle peak and slower reset times causes the ride side of the triangle to develop a slow tail toward zero error. This realization is particularly useful to switching to integrating tuning rules for near integrating processes and setting an arrest time that can be as small as 1/2 dead time for fastest load response.

Load Performance - Practical Limits from Tuning

Shinskey provided an insightful equation for integrated error for an unmeasured step disturbance. The numerator is simply the sum of the PID reset time, scan time, and filter time. The denominator is the controller gain. The change in PID output required is converted to an open loop error that puts the open loop gain in the denominator. An equation for peak error is possible starting with the one by Peter Harriott for a proportional only control and then adding an approximation of the slight effect of integral action.

Load Performance - Ultimate Limits from Dynamics

Often we lose sight of the ultimate limits imposed by process dynamics, particularly the total loop dead time. The peak error is the ratio of the loop dead time to the summation of the loop dead time and largest lag in the loop ideally set by a large back mixed process volume. The integrated error has the dead time squared in the numerator. These equations emphasize the importance of minimizing dead time.

Sources of Dynamics

Academic publications present a linear idealized view of process dynamics. The result is a loss of understanding of effect of automation system dynamics that for many loops limit the performance. These dynamics are the responsibility of the process control engineer and the ability to make improvements are faster and less expensive than changes to equipment and piping. The total loop dead time is the summation of all the pure delays and the fraction of time constants in series that become equivalent dead time in a first order plus dead time approximation. The largest time constant is hopefully in the process to slow down the process excursion from a load disturbance. However, for many liquid pressure and flow control loops, the largest time constant is in the automation system from valve or measurement response or process variable filter. A secondary time constant can be singled out to provide a second order approximation. The conversion of a near-integrating process gain to an integrating process gain is noted as simply the process gain divided by the process time constant. The open loop gain is the product of the valve or VFD gain, process gain and measurement gain showing the effect of valve and pump sizing and measurement scale. The resulting open loop gain is dimensionless for self-regulating processes and 1/sec for integrating processes. Many people do not realize that the PID works in percent of signals and not engineering units.

Dead time from Measurement System

There are many sources of the measurement dead time including transmitter update rate, analyzer cycle time, sensor lag(s), sensor transportation delay, transmitter damping, and resolution or sensitivity limit that is the limit divided by the process variable rate of change. The dead time from valve backlash and stiction can similarly be estimated as the dead band or resolution limit divided by the rate of change of controller output.

Use of Digital Twin to Become More Like Shinskey

We have incredible untapped capability in today's measurements, control valves, controllers, and software. What we are missing is the recognition of what Shinskey's knowledge and our systems can do to increase the performance of the operator and plant. The Digital Twin can reveal the opportunities, detail the solutions, and identify and achieve the benefits that can lead to a rejuvenation of our profession providing the motivation utilizing and building on what we learned from Shinskey. Detailed first principle dynamic models with automation system dynamics added can reveal the actual performance revealing the consequences of process and mechanical design, control strategies, procedure automation, and PID options and tuning. The powerful capability of new blocks and modeling objects enables the average process control engineer to fully exploit the opportunities and give the knowledge gained to all of the personnel running and supporting the plant. The Digital Twin has offers technical fluency and skill in addressing process control improvement opportunities in pH, chemical reactor, bioreactor, and compressor control. The key roles are emphasized in exploring=> discovering => prototyping => testing => justifying => deploying => testing => training => commissioning => maintaining => troubleshooting => auditing => continuous improving showing "before" and "after" benefits of solutions from online metrics. It is important to note that while we have singled out process control engineers in this presentation, the need for knowledge to attain the best performance extends to operators, maintenance technicians, process engineers, mechanical engineers, and information technology specialists (e.g., data scientists). Just think what can be realized if we were all on the same page understanding the process and operational opportunities and the value of the best measurements, valves, strategies, and PID?

Digital Twin's Key New Blocks

Blocks have added to the Digital Twin to test innovations and improvements in strategies, dynamics and PID tuning. The PID Performance block computes the integrated and peak error for load disturbances and the rise time, settling time, overshoot and undershoot for the setpoint response. The block is automatically triggered by a change in PID setpoint or output. A Backlash-Stiction block can be put in series with each other to simulate what is happening in the actuator, shaft to stem connection, and stem to ball or disk connection. Blocks to include a variable dead time, slewing rate and response lags can be inserted. The Future Value block can provide a timely low noise rate of change and future value typically a factor of the total loop dead time into the future from current value. This is extremely useful for batch endpoint prediction, profile control, procedure automation, surge curve and actual surge identification, RCRA pH violation prevention, and impending Safety Instrumented System activation.

Conclusion

This is just a brief insight into the incredible deep understanding gained and great advances achieved by Shinskey. Reference 2 below includes a list of nearly all of the articles and books by Shinskey. We should read and learn and use the wealth of knowledge given by Shinskey.

References

1. McMillan, G.K. and Vegas, P. H., *Process/Industrial Instruments and Controls Handbook Sixth Edition*, McGraw-Hill, 2019 https://www.mhprofessional.com/9781260117974-usa-process-industrial-instruments-and-controls-handbook-sixth-edition [mhprofessional.com]
2. McMillan, G.K. and Nino, S., "The greatest source of process control knowledge", *Control*, Oct 2017 https://www.controlglobal.com/articles/2017/the-greatest-source-of-process-control-knowledge/?stage=Live
3. McMillan, G.K. and Berutti M.E., "Virtual plant virtuosity", *Control*, April 2017 https://www.controlglobal.com/articles/2017/virtual-plant-virtuosity/?stage=Live
4. McMillan, G.K., "How to specify valves and positioners that don't compromise control", *Control*, May 2016 https://www.controlglobal.com/articles/2016/how-to-specify-valves-and-positioners-that-dont-compromise-control/?stage=Live
5. McMillan, G.K., *Tuning and Control Loop Performance, Fourth Edition*, Momentum Press, 2014 http://www.momentumpress.net/books/tuning-and-control-loop-performance-fourth-edition

551691 Importance of Regulatory Control Design for MPC

Greg Shinskey has had a tremendous impact on the practice of process control engineering going back to the late 1970s and 1980s. During this time frame, the arrival of the Distributed Control System (DCS) and the minicomputer allowed Shinskey's process-oriented, advanced regulatory controls to be readily implemented. Taking further advantage of increasing computing power, Model Predictive Control applications came on the industrial scene in the mid 1980s, with significant growth coming in the 1990's and thereafter. MPC provided built-in capability to handle process/control interactions, constraints, and optimization, ostensibly removing the need to improve or modify the regulatory design.

With the benefit of experience gained over many MPC applications, the recognition of the importance of the regulatory control design for MPC has only increased, and an acknowledged best practice is to jointly design both the MPC and regulatory controls. Design considerations include disturbance rejection, constraints, and optimization objectives. As a result, many of the techniques and reasoning behind Shinskey's approaches to process control hold true today.

In this presentation, we will highlight, via examples primarily involving distillation columns, the considerations that go into the design of the regulatory controls and MPC, while making reference to Shinskey's contributions.

551693 The Forgotten Science of Distillation Control

Distillation is the heart of all refining and petrochemical plants, as it determines the ability of a plant to

generate on specification product(s). In spite of the vast body of fundamental research and practical

design experience within the industry, the design and tuning of the regulatory controls, which plays a

pivotal role in achieving on specification production, became and remains largely a lost and dying art.

This is evidenced by the author's experience across multiple sites and companies wherein similar, if not

identical, controls design and control ability issues were discovered and rectified on and around

distillation units, as well as by the precipitous drop in attendance of control related sessions at industrial

and academic conferences.

Historically, there was a great amount of attention given to the design and performance of the

regulatory and advanced regulatory controls (ARC) implemented on distillation units. As more advanced

technologies came into common use, less attention was given to the underlying regulatory control

structures, in spite of the fact that it provides the foundation upon which the new technologies rely. IN

parallel to the shift in technology usage, the past decades have seen fundamental changes in the modus

operandi of refineries: the design and operations resemble that of chemical plants. This was primarily

the result of end user quality programs that propagated more stringent product specifications

"upstream" in the supply chain. Despite the changes to the operational criteria these tighter

specifications created, refinery designers have effectively maintained the historical "status quo",

especially regarding controls design.

With these technological and historical aspects in mind, this paper presents fundamental and practical

guidelines for designing robust and reliable control structures on these vital units. The authors draw from 30 years of experience in process and controls design, and tuning such controls, on multiple units

across a wide array of refining, petrochemical and LPG/LNG plants. In addition to the guidelines, a

sampling of real world situations and the solutions for same will be presented to illustrate how the

principles are practically implemented. Topics included in the presentation will include:

- Only Two "Handles" for Every Distillation Tower
- Accumulator Level Control Myths
- Relative Gain Analysis Made Easy
- "Closing the Loop" in Controls Design
- When Tray Temperatures act like Levels
- Flooded Condenser Design: Keep it Simple!
- APC: Love it, Hate it, Need it?

549967 Controlling Trace Impurities in a Dividing Wall Distillation Column

Though attractive due to capital and operational savings, dividing wall distillation columns (DWCs) have yet to gain widespread industrial acceptance. One notable concern is controllability. The research in this presentation investigated the operational flexibility of a fixed-design DWC using a four component mixture, tested the ability of conventional distillation control design techniques to determine control structures for a DWC, and created a minimum energy operating surface that could be used to examine control structures.

As a form of process intensification, DWCs have a higher risk for controller interaction rendering conventional PID control potentially inadequate. This work successfully used two PID temperature controllers to maintain the column at steady state, transition the column between steady states, and reject feed disturbances without controller interaction. These controller pairings were determined using conventional controller design techniques. Therefore, for this chemical system and column design, traditional approaches to distillation control are sufficient to handle the intensified nature of DWCs.

Because more components are present in DWCs in larger amounts, there is concern that temperature control will no longer imply composition control. Temperature control proved successful in this study. Controlling two temperatures maintained column operation against feed disturbances. In addition, prefractionator temperature correlated well with reboiler duty for multiple feed qualities therefore serving as a promising control variable. A steady state model with heat transfer matching the pilot data was scaled to the size of an industrial tower and used to generate a minimum energy response surface for different vapor and liquid split values.

551768 The Contingencies of Distillation Columns for Ers Design and Special Reference to the Location of the Relief Device

This presentation emphasizes the unique contingencies applicable in the emergency relief system design of tray and packed columns used for mass-transfer operations. Attention is given to the internals of the column and their impact on such phenomena as flooding and obstruction to the flow required to attenuate pressure during an emergency relief. While the relief devices should always be located at top of the column to promote unidirectional vapor relief flow and process vapor flow thereby avoiding reverse vapor flow of relief and consequential damage of internals, there are circumstances when the flow of the relief device(s) located at the top of the column is high enough to cause flooding. An unwise mode of differential pressure control with coincident double jeopardy of overhead condenser failure may also create high relief load. The flow may be obstructed due to softening of non-metallic packing material under a fire scenario, coking, or formation of pop-corn polymers. Too high a flow may cause a pressure drop through a tray that may damage the tray. Under these circumstances the choices are consideration of metal packing, redesign of tray, tray support, and an inherent means of reducing flow through relief device by restricting lift or restricting the scenario flow, safety instrumentation, or location of a relief device at the bottom section of the column in the vapor space. If the location of the relief device at the bottom of the column is unavoidable, two-phase flow relief along with phase separation in effluent handling equipment and safe relief discharge location must be considered. This is a price that a company must pay for irresponsible project and process management in not engaging an Emergency Relief System (ERS) design Engineer in the conceptual and design phases of the engineering.

In the end, an ERS design of a mass-transfer column without the consideration of the internal is not only hollow, but also equally irresponsible.

551769 A Practical and Cost-Effective Solution to Monitor Product Recovery Rate in High Purity Side Draw Distillation Columns: An Industrial Methanol Distillation Case Study

The measurement of product recovery ratio in industrial methanol distillation is of high economic importance and represents a key performance index (KPI) for the distillation unit. However, in many plants the product recovery of the distillation units is not actively monitored. Instead, it is back calculated based on daily production figures. This is primarily because the active monitoring of product recovery can be costly as this requires the installation of expensive gas chromatographs and accurate feed mass flow measurements, which can be difficult to justify economically. In this work we have developed a simple and economical method based on density and flow rate measurements to calculate the product recovery of industrial methanol distillation columns. This method has been validated against plant measurements as well as a validated process simulation. Step and disturbance tests carried out suggest the proposed method is able to accurately estimate the product recovery within the plant operational envelope.

544144 Multibatch Distillation Systems (MultiBatchDS™) and Novel Batch Configurations

Batch distillation is an unsteady-state operation. It is because of this unsteady-state nature that batch distillation columns can be operated in multiple configurations, multiple operating modes, and can have multiple fractions or cuts. The new column configurations like the batch stripper and the middle vessel columns can circumvent the steady-state mass balance restrictions of continuous columns resulting in better purity and better yields.

Multibatch distillation systems (**MultiBatchDS™**) is a software package which captures the essence of these multiple configurations, multiple operating modes, and various fractions of batch distillation by providing the capability to simulate, design, optimize, and control such columns with multiple levels of models.

This talk presents few real-world case studies different operating companies, demonstrating the usefulness of new column configurations of batch distillation.

551770 Details of Industrial Batch Distillation

Batch distillation is one of the oldest separation processes. It is still widely used in chemical industry due to its flexibility: one column can be used for multicomponent separations and for different mixtures.

Design of batch distillation is discussed, with special attention to thermodynamic properties of the mixture. Selection of process parameters (heat duty, reflux ratio, pressure and cut sizes) plays important role in process optimization. In non-ideal systems the products obtained from batch distillation may be counterintuitive; for example the less-volatile components can be obtained before the more-volatile species.

Typical operating steps depend on the feed mixture and distillation objectives (how many products need to be made and what is the desired purity). These steps may include charging, heating and degassing, establishing reflux, low-boils cut, front cut, pure cut, after cut, pumping out or washing the pot. Operational details of all these steps are discussed; for example the feed line size impacts the time of the batch; or flashing product tanks in the first batch is necessary to reach the purity specification. Most of these operational details are dictated by common sense but they are necessary for running successful batch distillation.

Some examples of plant trials are given.

550022 On the Random Packing Hydraulics for a Corrosive Application

The Chlor-alkali industry uses electrolytic processes to produce chlorine and caustic soda from aqueous sodium chloride solutions. Chlorine drying is one of the stages of chlorine processing which removes moisture from chlorine by using sulfuric acid. Two or more columns are employed in series for chlorine drying by counter-current contact of gas flow with acid flow in a packed column. For the latter, material selection is pivotal to the process design because even small traces of water in chlorine can be highly corrosive. Water content in chlorine and sulfuric acid drive the material selection, which in turn is dependent on each material's chemical resistance, temperature rating, mechanical strength and cost. Design and economic considerations that lead to selecting materials for the column shell, internals and high capacity random packing are discussed in this paper. Quantitative comparisons of the different packing materials in terms of their hydraulic properties will be shown. This will be supported using select commercial application examples.

550226 Necessity of Curvature in Distillation Boundary for Feasible Extractive Distillation Systems in Separating Maximum-Boiling Azeotropes

Extractive distillation is commonly used in industry for separation of azeotropic mixture. A heavy entrainer is added into an extractive distillation column to greatly enhance the relative volatility of one component so that it can go to the distillate of this column while the entrainer and the other component go to the column bottoms. These two components can easily be separated in an entrainer recovery column with heavy entrainer at bottom stream for recycling to the first extractive distillation column. For the separation of maximum-boiling azeotropes, design configuration of the two-column system is exactly the same as the ones for minimum-boiling azeotropic systems. However, because of different topology of ternary diagram for the maximum-boiling system, unique design consideration should be taken to properly design such separation system. Unlike minimum-boiling system having no distillation boundary in the ternary system, there will be a distillation boundary for the maximum-boiling system with a heavy entrainer. Depending on the degree of curvature of distillation boundary, feasible minimum entranier-to-feed ratio can be estimated with given product purity specifications. This minimum ratio can be considered as another screening tool for selecting suitable entrainer of maximum-boiling azeotropic systems. Two examples are presented to demonstrate the important findings for the separation of maximum-boiling azeotrope via extractive distillation system. In the first example of separating acetone and chloroform, over 20% savings in total annual cost and energy can be realized by using a newly proposed entrainer of N-methyl-2-pyrrolidone for this system as opposed to using literature suggested dimethyl sulfoxide or ethylene glycol as entrainer. In the second example of separating phenol and cyclohexanone, the savings in total annual cost and energy are even more significant by using a newly proposed entrainer for this system as opposed to using literature suggested acetophenone adding into this separation system.

Neural Network based Soft Sensor for Pilot-Plant Distillation Column

Thaer A. Abdulla[1], Mohanad Osman[2] and Thomas F. Edgar[1]
(1)The University of Texas at Austin, Austin, TX, US, (2)Canadian Natural Resources Limited, Calgary, AB, Canada

Abstract

The development of an inferential soft sensor for a pilot-plant distillation column separating an ethanol-water mixture using neural network (NN) method has been investigated in this work. Inferential sensors are increasingly used in the process industries to infer the value of the main quality variable while utilizing much easier to measure secondary variables of the process. The lags between the input variables and the output variable vary due to changes in operating conditions. Previous studies have introduced different methods to estimate lags for input and output variables, but all of them have assumed these lags to be constant regardless of the changes in the operating conditions. In this work, an inferential sensor that can infer the composition of ethanol at the top product using time lags for the input variables and varied first-order time constant lags with the output variable is developed. The developed inferential sensor is based on a neural network (NN) model. This NN model is built in MATLAB/Simulink. An OPC link is established in order to communicate between MATLAB/Simulink and the Emerson DetlaV Distributed Control System (DCS) of the pilot-plant distillation column. Principal Component Analysis (PCA) and Projection to Latent Structures (PLS) methods are used in this work to remove the outliers from the input variables set and to determine the most correlated values of the input variables and their lags with the output variable Xa (ethanol mole fraction of distillate product) respectively. The model adaptively selects the correct first-order time constant lags of an output variable according to the instantaneous operating condition (the composition of ethanol is increased or decreased) and assigns a best value for each case. The prediction performance of the proposed NN models (with and without time lags for input variables) is illustrated using experimental data from the pilot-scale ethanol-water distillation column. More than 400 samples were collected to create and validate the results of NN models. The proposed NN model with time lags for input variables and varied first-order time constant lags for output variable gave lower error (RSME=0.01) compared with the NN model without any time lag for input and output variables (RSME=0.06). In summary, a high accuracy soft sensor for the ethanol composition of the top distillation product has been developed and validated. Based on this soft sensor, an inferential PI controller and Model Predictive Controller (MPC) for this pilot-plant column will be developed in future work.

I. Introduction

Recent advancements in measurement techniques in the process industries and increasing plant data collection and storage capabilities have encouraged the trend to estimate some important but difficult to measure process variables from the data collected of other easy to measure variables. The mathematical models used for this estimation are called Soft Sensors or Inferential Sensors. Inferential sensors can be classified into first principles model (FPM) inferential sensors and data-driven inferential sensors. An FPM inferential sensor is constructed by rigorous mathematical derivation of the model from the first principles chemical and physical equations that describe the process. Example of FPM inferential sensors can be found in [1]. FPM inferential sensors suffer from two main drawbacks: high computational cost due to complexity and model inaccuracies due to the simplification assumptions required to solve the first principles equations. Consequently data-driven inferential sensors which are constructed based purely on the process data without the need of knowledge of the process fundamentals are gaining more attention. A number of hybrid inferential sensors based on combinations of the two types have also been introduced [2, 3, 4]. Generally, these hybrid inferential sensors use a data-driven regression technique to estimate some of the parameters in the first principles equations that are two complicated to be driven mathematically. A comprehensive review on inferential sensors can be found in [5, 6].

Generally, data-driven inferential sensors are developed through the following steps: (1) Data pre-processing, (2) Selection of model building technique, (3) Selection of important process variables and their lags, and (4) Model building/validation. The first three steps will be explained in more detail below.

Data pre-processing step mainly includes removal of outliers and data normalization. Sensor faults, signal noise, human errors and unusual operating conditions can cause variables to have unjustifiable extreme values in some observations, these observations are labeled outliers [7]. Including outliers during the construction of the inferential sensor can deteriorate its performance; hence their removal from the data set is essential. In some cases, outliers can be identified via human practical knowledge of the process. Data during unrepresentative operation periods (e.g., start up, shut down) or considered to by the plant experts to be having unusual values based on engineering judgment are labeled outliers and excluded. A more common practice is to identify outliers by the help of statistical analysis of the data {median absolute difference (MAD) [8], principal component analysis (PCA) [9], and Hotelling's T^2 statistics [10]}. Different plant data variables usually have significantly different ranges of their numerical values and hence significantly different variances. For the variables with higher variance to not have disproportionate influence in the model construction data, normalization (or scaling) is crucial. There are several methods of normalization: min-max normalization, unit-variance normalization and zero-mean unit-variance normalization [9].

Selection of model construction technique will depend on many factors; some of them are the degree of process nonlinearity, level of variables collinearity, the importance of the multivariate aspect of the process, number of process variables, the amount of the collected data, level of noise, degree of knowledge of process fundamentals and dynamics and the required state of tradeoff between model accuracy and complexity. The most common techniques used in the construction of inferential sensors are multivariate analysis techniques (PCA and PLS [11, 12], NPCA and NPLS [13, 14], and DPLS [9]), neuro-fuzzy (NF) [15, 16], and artificial neural networks (ANN) [17, 18, 19]. The development of ANN and its applications in chemical engineering can be seen in [20]. The performance of these models can be measured by several indexes and one of the most common indices is the root square mean error (RSME). The model with lower RSME will have the highest accuracy.

Finally, selection of input variables and their lags is very important in the soft sensor development process. Using redundant and collinear input variables increases the complexity of the model and has negative effects on the performance of the inferential sensor. Hence, selecting only a subset of the input variables that have the least co-linearity and the most correlation with the output is an important step of developing an inferential sensor [21]. Typically, the choice of the important variables and their corresponding lags is done simultaneously. The choice of the correct delays for the input variables has a high effect on the quality of the inferential sensor [22]. If the correlation between different delayed versions of an input with the output is studied, the input version with the correct delay should have the highest correlation value with the output. Based on this theory the linear correlation between each of the inputs in different models and the output is used to identify the correct lag for each input variable in [23]. One of the problems of this method is that when the value of the correlation of each input variable with the output variable is calculated individually, it ignores the effect of the other input variables. To solve this problem, Hoskins and Himmelblau [17] have suggested the use of the partial correlation as the identification criteria of the correct lag. Both methods can detect only the linear correlation and fail to work if the relations between the inputs and the output are nonlinear. A method that is applicable for both linear and nonlinear models was proposed in [24] based on Mallows' statistics that use a standardized total square error as a criterion for the selection of the inputs. Another method for nonlinear models, based on the computation of Lipschitz quotient was proposed in [25]. The last two methods suffer from the problem of high computational load. A much simpler method based on partial least squares (PLS) regression technique was proposed in [26].

II. METHODOLOGY

A- ANN Development

Two ANN models are trained for the prediction of ethanol mole fraction of the distillate product (Xa), first model has input lags and the second without input lags. Both NN models are Feed Forward (FF) networks. Levenberg-Marquardt back-propagation with momentum technique is used as the training algorithm for both networks. Trial and error approach is used to find the optimum number of hidden layers, number of neurons of each hidden layer and the activation functions. The network structure of both models (as shown in Table 1) that has the minimum RMSE on the test set is selected for the inferential sensor. Based on this, the structure of four layers (three hidden layers and an output layer) with the neurons (5, 3, 3) and (10, 7, 3) for the network with input lags and the one without input lags respectively has the best performance (RSME=0.02 and 0.04 for networks with input lags and without lags respectively). The activation function of the first hidden layer of both networks was Log-Sigmoid activation function (logsig) and pure-linear activation function (purelin) for other layers of both networks. The description of these networks can be seen in Figures 4 and 5 and Table 1.

In this work, zero-mean unit-variance normalization method has been used for data normalization and PCA method is used for removing the outliers from the input variables set as shown in Figure 1. A PLS method was used to determine the most correlated values of the input variables with the output variable (ethanol mole fraction) (as shown in Figure 2) and the most correlated values of the input lags with the output variable (ethanol mole fraction) (as shown in Figure 3). The variable influence plot (VIP) represents a weighted sum of squares of the weights (WSSW) assigned to the variables in the PLS model, which summarizes the influence and thus the importance of each variable in the prediction of the primary output. Figures 2 and 3 show the VIP of the variables and their lagged versions respectively. The highest WSSW value for the input variables with their lags is selected. For each of the selected input variables, a sequence of 5 lags (k-2, k-4, k-6, k-8, k-10) was obtained with two sample delay between each other, then the variables with their lagged versions were used in the PLS model. Based on these criteria, the following input variables with their lags were selected [4 min lags for top vapor temperature (T200), tray 2 temperature (T102), and tray 3 temperature (T103) were selected while 2 min lag for top column pressure (P081) was selected].

B- Equipment and Experimental Work

The equipment studied in this work is a pilot-scale distillation column of ethanol-water system. The process flow diagram (PFD) of this column can be seen in Figure 6. The column is capable to work either in batch or continuous operations. It has multi input ports to each tray in addition to the main feed entrance to the horizontal reboiler. There are 2 windows at the two bottom trays to watch the column behavior during the operation. The column is 6-inch inner diameter with 10-inch tray spacing, consisting of 10 sieve trays, condenser, reflux drum, and reboiler, and works under atmospheric pressure. The horizontal reboiler is initially filled with feed (20-22 gallon of 20-25 wt% of ethanol). Tables 2 and 3 show the design specifications and the nominal operating conditions of the column. It is desired to develop an inferential sensor that can accurately predict the mole fraction of ethanol (XA) at the top distillate product (the output) using different input variables of the process. The DCS control system (DeltaV) is used for controlling the column operation. The descriptions of the process variables tags can be seen in Table 6.

The column pressure is controlled using nitrogen. There are two valves, one for nitrogen supply and the second for the vents of the column. The PID pressure controller determines which valve will be opened. The subcooled condensate temperature (45 °C) is controlled by the water-cooling flowrate. The reflux drum level is controlled by the flow rates of reflux and distillate streams. Reboiler steam, feed, product, and reflux flow rates are controlled using PID controllers. The reflux drum level controller is usually at Auto mode (Set Point=10 inch) while either the reflux flow controller will be in cascade mode (constant distillate flow operation, D=0.35 l/min) or distillate flow controller will be in cascade mode (constant reflux flow operation, R=0.75 l/min) or both of them will be in cascade mode (constant reflux ratio, R=3).

For the development of the inferential sensor, data were generated using experimental tests. The column was operated at different operating conditions [feed amount (70-85 liter) and feed composition (0.2-0.3 mass fraction of ethanol)] as shown in Table 4. A series of step changes in the reflux ratio (0.5-5) and reboiler steam flow rate (0.55-0.95 kg/min) are introduced as excitation signals to produce a wide range of temperature profiles and top product concentrations. Samples from the distillate product stream are collected regularly at fixed interval (5 to 6 min). The process (input/output) variables are stored within DeltaV (DeltaV history collection) and can be exported to excel sheet at a sampling interval (10 sec or more). Around 400 samples were collected with their corresponding operating conditions. These samples are segmented into two subsets: 160 samples training and testing sets and 240 samples validating set. The RSME of both networks during the validation step was (0.03 for model with input lags and 0.06 for model without input lags, see Figure 7).

C- Output Delay Implementation

Most of the previous works of NN application for distillation columns {estimation and simulation applications [27, 28], optimization applications [29, 30], and control applications [31, 32]) did not account for the whole dynamic path (dynamic behavior) of the output variables (product compositions) and focused only on the final steady state of the output variables so they either did not use any time constant delay for the outputs or just used a constant delay for the outputs. In order to improve the accuracy of our soft sensor during the whole dynamic path of the ethanol mole fraction (output variable) so it will be useful for the inferential control purpose, a variable (updated) time constant delay for the output is used in addition to the constant lags for the input variables. From our experimental investigations, it takes more time (slow response) when we want to increase the composition of ethanol (in other words, long dynamic path of the ethanol composition until reaching the final steady state from below) while it takes less time (fast response) when we want to decrease the ethanol composition (in other words, short dynamic path of the ethanol composition until reaching the final steady state from above). So, several pairs of first-order time constant lags are used for both cases (ethanol composition is increased/deceased). A past value for the ethanol composition (0.5 min earlier) is subtracted from the current value of ethanol composition. If the result of the subtraction is positive (ethanol composition is increased), those first-order time constant lags (3 min, 5min, 7min, and 9min) are tried and if the result of the subtraction is negative (ethanol composition is going down), those first-order time constant lags (1min, 3 min, and 5min) are tried. So, the following pairs of first-order time constant lags {(1min, 3min), (1min, 5min), (1min, 7min), (3min, 3min), (3min, 5min), (3min, 7min), (5min, 5min), (5min, 7min), (5min, 9min); note the first value of these pairs is when the composition is going down while the second value is when the composition is going up} are used to create different sensors (AI91E, AI91EPV1 , AI91EPV2, AI91EPV3, AI91EPV4, AI91EPV5, AI91EPV6, AI91EPV7, AI91EPV8) as shown in Table 5 .

III. RESULTS & DISCUSSION

A- Initial Validation Results of the Soft Sensor

Figure 7 shows a comparison between the performance of the ANN inferential sensor model with input lags and the ANN inferential sensor model without input lags for the online prediction based on the validation set. It is clear that the inferential sensor model with input lags has significantly outperformed the one without lags, achieving a considerably lower RMSE of (0.03).

B- Improved Validation Results of the Soft Sensor

Online tests were conducted for the proposed adaptive selection of time constant lags inferential sensor. It is clear that the varying output lags inferential sensor have surpassed the ones with only constant input lags or without input lags in the prediction accuracy in all these tests as shown in Table 5 and Figure 8. The sensor with best time constant lag configuration (3min, 7min) for the output variable (ethanol mole fraction) in addition to the constant input lags of the model has been selected (AI91EPV5) in our work so the accuracy of this sensor has been increased (RSME=0.01).

IV. CONCLUSIONS AND DIRECTIONS FOR FUTURE WORK

In this work, it was found that the NN model with input variable lags has more accuracy compared with the NN model without input variable lags. The time constant lags of the output variable (ethanol mole fraction of top product) are significantly dependent on the instantaneous value of the operating conditions (whether the ethanol composition is increased or decreased). The NN model with the updated time constant lags for output variable in addition to the input variables lags developed in this work was able to account for the whole dynamic path (dynamic behavior) of the output variables (product composition). This model is found to outperform a model with the input variables lags or the one without input variables lags. The high accuracy of the inferential sensor in tracking the value of the actual top product composition closely made it possible for an effective control scheme to be applied on the system. Based on this inferential sensor, inferential PI controller and MPC (DeltaV MPC/Aspen DMCplus) will be developed and their performance will be compared for future work.

V. REFERENCES

[1] Luo, R.F., Shao H.H., Zhang Z.J. (1995), Fuzzy-Neural-Net-Based Inferential Control for A High-Purity Distillation Column, Control Eng. Practice, Vol. 3, No. 1, pp. 31-40

[2] Gnoth, S., Jenzsch, M., Simutis, R., Lübbert, A. (2008), Product formation kinetics in genetically modified E. coli bacteria: inclusion body formation, Bioprocess Biosyst Eng, Vol. 31, pp. 41–46

[3] Kim, B. J., Kim, I. K. (2005), An Application of Hybrid Least Squares Support Vector Machine to Environmental Process Modeling1, Parallel and Distributed Computing: Applications and Technologies, Lecture Notes in Computer Science, Volume 3320, pp. 57-85

[4] Fortuna, L., Grazaini, S., Rizzo, A., Xibilia, M. G. (2007). Soft Sensors for Monitoring and Control of Industrial Processes. Springer-Verlag, London.

[5] Kadleca, P., Gabrys, B., Strandtb, S. (2009), Data-Driven Soft Sensors in The Process Industry, Computers and Chemical Engineering, Vol. 33, No. 4, pp. 795-814

[6] Lin, B., Recke, B., Knudsen, J. K. H., Jørgensen, S. B. (2007), A systematic approach for soft sensor development, Computers and Chemical Engineering, Vol. 31, pp. 419–425

[7] Shu Xu, Bo Lu, Michael Baldea, Thomas F. Edgar, Willy Wojsznis, Terrence Blevins, Mark Nixon, Data cleaning in the process industries, Rev Chem Eng 2015; 31(5): 453–490

[8] Liu, J., Min, K., Flan C., Chang, K. S. (2000), Robust Nonlinear PLS Based on Neural Networks and Application to

Composition Estimator for High-Purity Distillation Columns, Korean J. Chem. Eng., Vol. 17, No. 2, pp. 184-192

[9] Eriksson, L., Johansson, E., Kettaneh-Wold, N., and Wold, S. (2001). Multi- and Megavariate Data Analysis. Umetrics Academy, Sweden.

[10] Shin, J., Park, S. (2000), Design of a Composition Estimator for Inferential Control of High-Purity Distillation Columns, Chemical Engineering Communications, Vol. 178, No. 1, pp. 221 - 248

[11] Fujiwara, K., Kano, M., Hasebe, S. (2008) Correlation-Based Just-In-Time Modeling for Soft-Sensor Design, Computer Aided Chemical Engineering, Vol. 25, pp. 471-476

[12] Zamprogna, E., Barolo, M., Seborg, D. E. (2005). Optimal selection of soft sensor inputs for batch distillation columns using principal component analysis. Journal of Process Control, Vol. 15, pp. 39-52.

[13] Mark A. Kramer, Nonlinear Principal Component Analysis Using Autoassociative Neural Networks, AICHE Journal, Vol. 37, No. 2, February 1991

[14] E. C. Malthouse, A. C. Tamhane and R. S. H. Mah, Nonlinear partial least squares, Computers chem. Engng Vol. 21, No. 8. pp. 875-890, 1997

[15] Arauzo-Bravo, M. J., Cano-Izquierdo, J. M., Gomez-Sanchez, E., Lopez-Nieto , M. J., Dimitriadis, Y. A., Lopez-Coronado, J. (2004), Automatization of a penicillin production process with soft sensors and an adaptive controller based on neuro-fuzzy systems, Control Eng. Pract., vol. 12, pp. 1073–1090

[16] de Canetea, J. F., del Saz-Orozcoa, P., Gonzalez-Pereza, S. (2008), Application of Adaptive Neurofuzzy Control Using Soft Sensors to Continuous Distillation, Computer Aided Chemical Engineering, Vol. 25, pp. 465-470

[17] Hoskins, J. C., Himmelblau, D. M. (1988), Artificial Neural Network Models of Knowledge Representation in Chemical Engineering, Comput. Chem. Eng., Vol. 12, No. 9/10, pp. 881-890

[18] Gonzaga, J.C.B., Meleiro, L.A.C., Kiang, C., Filho, R. M. (2009), ANN-based soft-sensor for real-time process monitoring and control of an industrial polymerization process, Computers and Chemical Engineering, Vol. 33, No. 1, pp. 43–49

[19] Wang, X., Ding, Y., Shao, H. (1998), the Improved Radial Basis Function Neural Network and its Application, Artificial Life and Robotics, Vol. 2, No. 1, pp. 8-11

[20] Venkat Venkatasubramanian (2019), The Promise of Artificial Intelligence inChemical Engineering: Is It Here, Finally?, AIChE Journal, Vol. 65, No. 2, pp. 466-478

[21] Feng, R., Shen, W., Shao, H. (2003), A Soft Sensor Modeling Approach Using Support Vector Machines, Proceedings of the 2003 American Control Conference, Denver, Colorado, June 4-6, Vol. 5, pp. 3702-3707

[22] Lin, B., Recke, B., Knudsen, J. K.H., Jørgensen, S. B. (2007), a Systematic Approach for Soft Sensor Development, Computers and Chemical Engineering, Vol. 31, No. 5, pp. 419-425

[23] Zamprogna, E., Barolo, M., Seborg, D. E. (2005), Optimal Selection of Soft Sensor Inputs for Batch Distillation Columns Using Principal Component Analysis, Journal of Process Control, Vol. 15, No. 1, pp. 39-52.

[24] Gomm, J. B., Evans, J. T., Williams, D. (1997), Development and Performance of a Neural-Network Predictive Controller, Control Engineering Practice, Vol. 5, No. 1, pp. 49-59

[25] Fortuna, L., Graziani, S., Xibilia, M. G., Napoli, G. (2005), Soft sensors for product quality monitoring in debutanizer distillation column, Control Engineering Practice, Vol. 13, pp. 499-508

[26] Warne, K., Prasad, G., Rezvani, S., Maguire, L. (2004), Statistical and Computational Intelligence Techniques for Inferential Model Development: a Comparative Evaluation and Novel Proposition for Fusion, Engineering Applications of Artificial Intelligence, Vol. 17, No. 8, pp. 871- 885

[27] Vijander Singh, Indra Gupta, H.O. Gupta (2007), ANN-based estimator for distillation using Levenberg–Marquardt approach, Engineering Applications of Artificial Intelligence, Vol. 20, pp. 249–259

[28] M.T. Vafaei, R. Eslamloueyan, Sh. Ayatollahi (2009), Simulation of steam distillation process using neural networks, chemical engineering research and design, Vol. 8 7, pp. 997–1002

[29] Funmilayo N. Osuolale, Jie Zhang (2016), Energy efficiency optimization for distillation column using artificial neural network models, Energy, Vol. 106, pp. 562-578

[30] Funmilayo N. Osuolalea, Jie Zhanga (2017), Thermodynamic optimization of atmospheric distillation unit, Computers and Chemical Engineering, Vol. 103, pp. 201–209

[31] P. Dutta, R. Russell Rhinehart (1999), Application of neural network control to distillation and an experimental comparison with other advanced controllers, ISA Transactions, Vol. 38, pp. 251-278

[32] Neha Sharma, Kailash Singh (2012), Model predictive control and neural network predictive control of TAME reactive distillation column, Chemical Engineering and Processing, Vol. 59, pp. 9– 2

VI. APPENDICES

Appendix (A): Figures

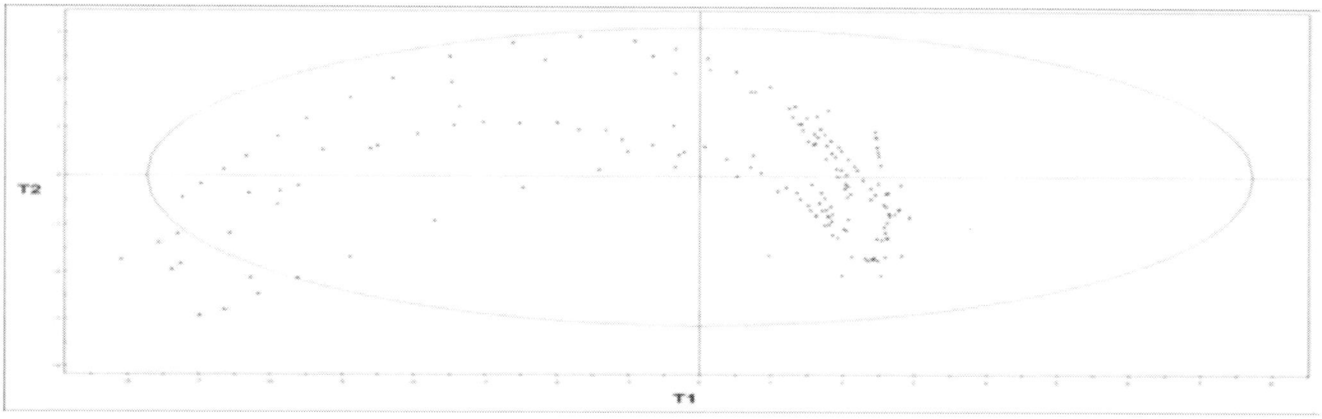

Figure 1 Score plot of the first two principal components of a PCA study for 169 observations

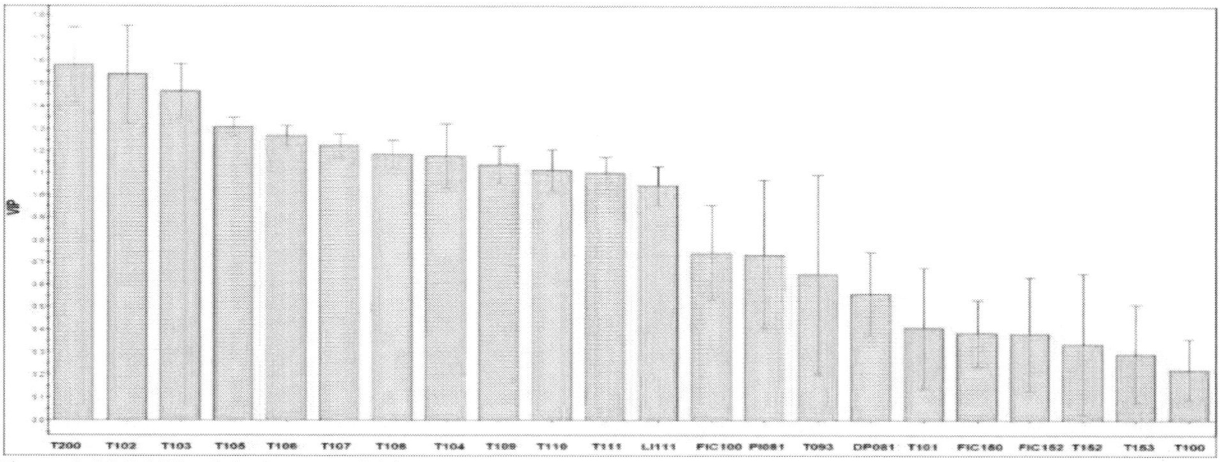

Figure 2 PLS-VIP of the input variables

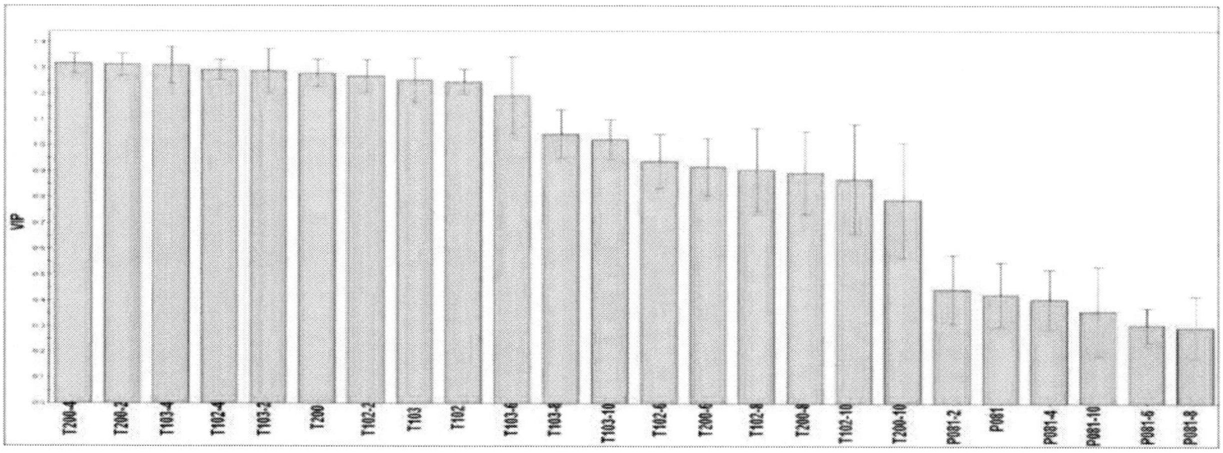

Figure 3 PLS-VIP of the input variables with their lags

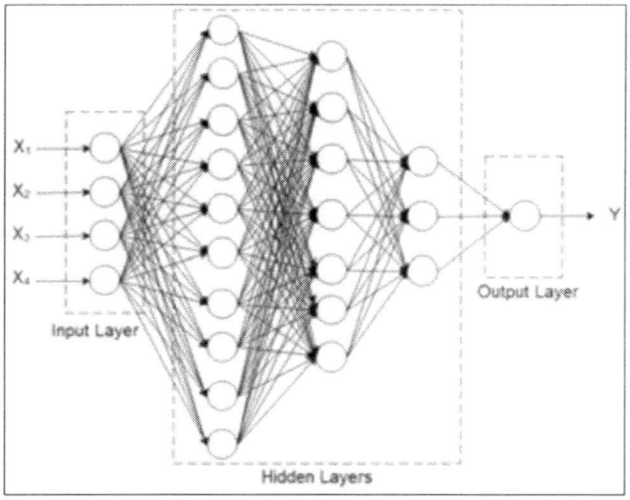

Figure 4 Structure of the NN Model without Input Delays

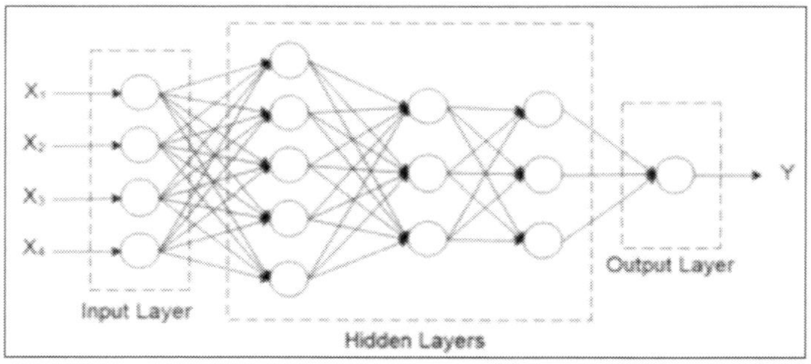

Figure 5 Structure of the NN Model with Input Delays

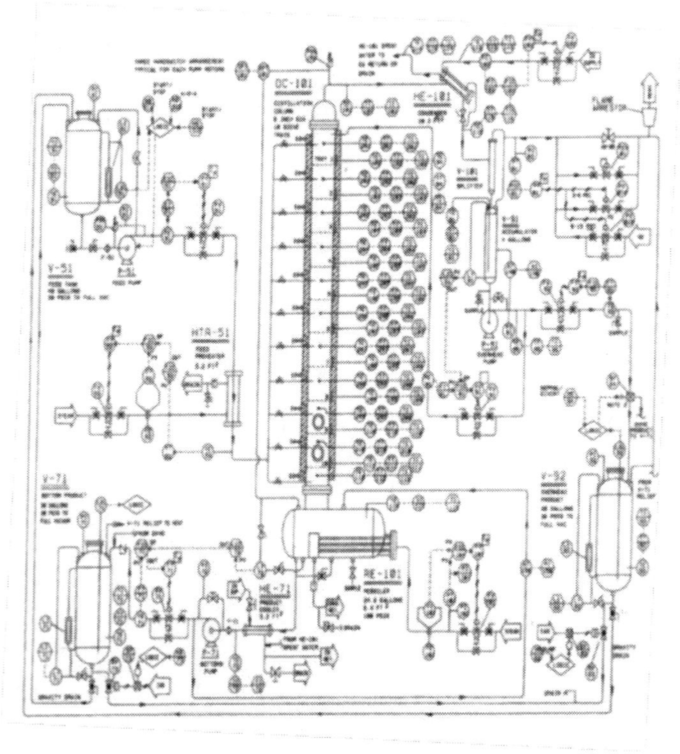

Figure 6 PFD of Ethanol-Water Distillation Column

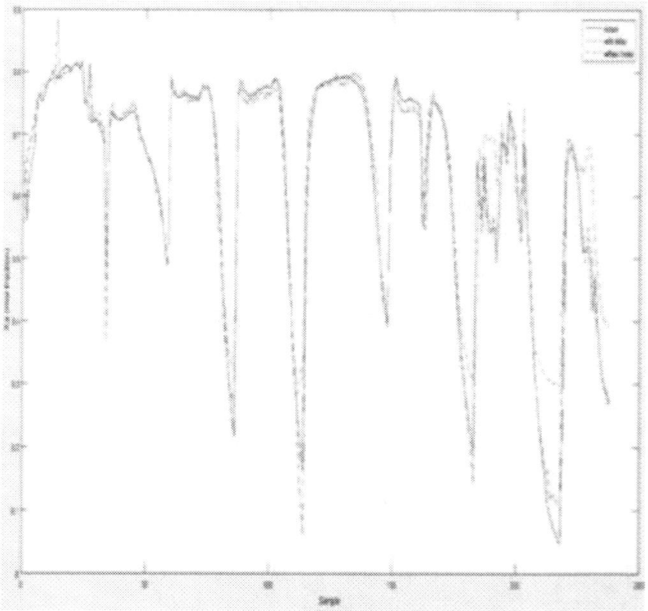

Figure 7 NN Soft Sensor Initial Validation Results

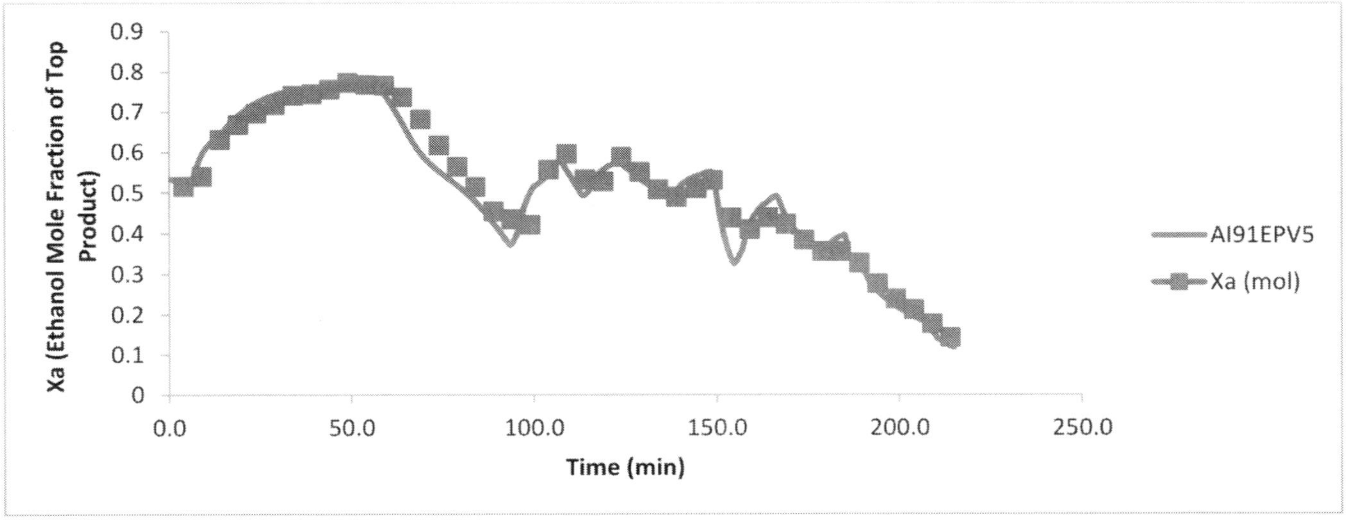

Figure 8 Improved NN Soft Sensor Validation Results

Appendix (B): Tables

Table 1 Neural Networks Structure

Name	Type	Training Algorithm	No. of hidden Layers	No. of Neurons in each hidden layer	Activation Function in the hidden and output layers
NN without input lags	Feed Forward	Levenberg-Marquardt back-propagation with momentum technique	3	10, 7, 3	Logsig, Purelin, Purelin, Purelin
NN with input lags	Feed Forward	Levenberg-Marquardt back-propagation with momentum technique	3	5, 3, 3	Logsig, Purelin, Purelin, Purelin

Table 2 Ethanol-Water Batch Distillation Column Specifications

Sieve trays specifications	
Number of sieve trays	10
Diameter (inch)	6.065
Tray space (inch)	10
Weir height (inch)	0.5
Lw/D	0.725
Fractional Active Area (Aa/A)	0.7
Fractional Hole Area (Ah/Aa)	0.29
Discharge Coefficient	0.8
Sieve Tray Hole Diameter (inch)	0.03125
Downcomer Area Fraction (Ac/A)	0.0278
Reboiler Pot Specifications	
Geometry	Horizontal Cylinder with Elliptical Heads
Length (inch)	14
Diameter (inch)	33
Heating Transfer Medium	Steam

Contd.	
Steam pressure (bar)	2
Steam Mass Rate (Kg/min)	0.8 kg/min
Steam Coil Configuration	
Bottom height (inch)	0
Top height (inch)	4.25
Area (m2)	0.596
Condenser Specifications	
Condenser Type	Shell and tube (2 pass design, city water on tubes, process condensate on shell side), Total Condensation with subcooling
Subcooled temperature (°C)	45
Condenser exchange area (m2)	1
Reflux Drum Specifications	
Geometry	Vertical cylinder with elliptical heads
Height (inch)	14
Diameter (inch)	6.5
Product tank Specification	
Geometry	Vertical cylinder with elliptical heads
Height (inch)	36
Diameter (inch)	14

Table 3 Water-Ethanol Batch Distillation Nominal Operation Conditions

Operation Conditions	
Feed Initial Charge (Gal or Lit)	20-22 Gal or 75-85 Lit
Feed Initial Composition (%wt)	18-28 %wt ethanol
Initial charge conditions (Temp and Pressure)	25 °C and 101 Kpa
Reflux Ratio, Constant Distillate, OR Constant Xd	3, 0.35l/min, Or 91%wt ethanol
Stopping criteria	Final ethanol mass fraction at Pot (Xw=0.005)
Top pressure (kpa)	101 Kpa
Reboiler Duty	0.8 kg/min sat. steam @3 bar

Table 4 Range of Operating Conditions used for the Training of NN

Operating parameter	**Range**
Reboiler steam flow rate (kg/min)	0.55 - 0.95
Distillate flow rate (l/min)	0.20 - 0.80
Reflux flow rate (l/min)	0.20 - 0.80
Reflux Ratio	0.50 - 50
Initial Feed Amount (Gallon)	17.50 - 22.50
Initial Ethanol Mass Fraction at Feed	0.18 - 0.28

Table 5 Soft Sensors Performance

Sensor	Model Type	RSME
AI91	NN without Input Lags	0.04013
AI91B	NN with Input Lags	0.02239
AI91E	NN with input lags and output time constants (1min, 3min)	0.01898
AI91EPV1	NN with input lags and output time constants (1min, 5min)	0.01897
AI91EPV2	NN with input lags and output time constants (1min, 7min)	0.01897
AI91EPV3	NN with input lags and output time constants (3min, 3min)	0.01746
AI91EPV4	NN with input lags and output time constants (3min, 5min)	0.01750
AI91EPV5	**NN with input lags and output time constants (3min, 7min)**	**0.01710**
AI91EPV6	NN with input lags and output time constants (5min, 5min)	0.02280
AI91EPV7	NN with input lags and output time constants (5min, 7min)	0.02313
AI91EPV8	NN with input lags and output time constants (5min, 9min)	0.02334

Table 6 Process Variables Tags Descriptions

Tags	Descriptions
T200	top column vapor temperature
T093	Subcooled condensate liquid temperature (°C)
T100	reflux temperature
T101	tray 1 (top tray) liquid temperature (°C)
T102	tray 2 liquid temperature (°C)
T103	tray 3 liquid temperature (°C)
T104	tray 4 liquid temperature (°C)
T105	tray 5 liquid temperature (°C)
T106	tray 6 liquid temperature (°C)
T107	tray 7 liquid temperature (°C)
T108	tray 8 liquid temperature (°C)
T109	tray 9 liquid temperature (°C)
T110	tray 10 (bottom tray) liquid temperature (°C)
T111	reboiler liquid temperature (°C)
LI111	reboiler level (cm)
T152	inlet water cooling temperature (°C)
T153	outlet water cooling temperature (°C)
FIC152	water cooling flowrate (l/min)
FIC100	reflux flowrate (l/min)
FIC150	steam mass flowrate (kg/min)
DP081	column differential pressure (kpa)
PI081	top column pressure (kpa)

544040 Molecule Management:Crude Unit Diesel Recovery

Diesel recovery continues as an important driver in refinery crude unit economics. Overall refinery profits are best when the desired molecule is recovered as soon as possible. Conventional refinery units recover diesel from the atmospheric column. Less commonly, vacuum towers can recovery a diesel product. However, these are not the only options.

Other configurations may include:

- gas oil towers,
- vacuum preflash towers, and
- gas oil splitters.

In addition, recycles between towers and improved stripping of diesel from gas oil sidestrippers can improve diesel recovery.

This paper:

- reviews the basic options (atmospheric and vacuum towers) and
- compares the basic options to the less common configurations.

Results are shown for an example crude for both distillation and cloud point specifications.

The analysis focuses on combining both the fundamentals of fractionation with an understanding of equipment limits and performance. Fundamentals covered will include system pressure and relative volatility and liquid-to-vapor ratios in the different options.

548745 Upgrading a Vacuum Tower at a Major South East Refinery

Many existing vacuum towers have been revamped to boost the capacity and or to increase the cut point more than once since their first start up. In most of the cases, only the mass transfer components of the vacuum column were upgraded. Fractionation trays were replaced with random packing, and subsequently with structured packing to minimize the pressure drop and thus the pressure at the flash zone with the final goals of maximizing throughput and distillates recovery.

Indeed rarely these revamps have addressed also a proper check of the fired heater, the transfer line, and the feed entry lay out to the flash zone of the vacuum tower. The increased throughput and reduced pressure will cause an increasing of the volume while the feed is traveling from the furnace coil outlet to the flash zone of the tower. A proper check of the velocity profile through the entire transfer line is mandatory to avoid the draw back associated to the risk of approaching the critical velocity at which, a consistent reduction of distillates yields may occur, or even mechanical damages.

This paper will describe the revamp of a vacuum tower at a major south east refinery where the transfer line of the fired heater was re-routed as the feed nozzle to the vacuum tower was relocated from tangential to radial. A detailed comparison of the operating data before and after the revamp will be also shown.

551231 Troubleshooting & Optimizing Refining Vacuum Towers

Vacuum column is a critical unit operation of crude unit. Two case studies will be presented.

Case Study #1: Tower pressure is a crucial parameter for optimum VDU column performance.

Operating pressure of a VDU column located in Europe was slowly rising, making vacuum residue quality unacceptable for bitumen production. Client suspected vacuum system efficiency was decreasing and planned for unit shutdown. In order to shorten shutdown duration, GTI Solutions (GTIS) was invited to perform a plant survey and to identify the faulty equipment.

GTIS not only identified the root cause but also provided solutions without a plant shutdown, saving client time and money.

This study demonstrates how a methodological troubleshooting exercise can help plant operators determine the root causes of VDU column malperformance and implement correct solutions.

Case Study #2: A refinery in Japan processing a mixture of Middle East crudes had experienced coking in the wash bed of a vacuum tower. Product qualities had gradually reduced whereas pressure drops increased over time. This occurrence repeated between plant turnarounds with varying degree of run-lengths and product qualities. As a corrective measure, the refinery was tasked to identify optimal packings for the wash oil bed.

This case study demonstrates the importance of packing selection, especially for the wash bed of a vacuum tower. Results from a series of plant turnarounds, where the wash bed was packed with various packing textures, will be presented.

551700 Modified Shed Decks Tackle Crude Tower Fouling Problems

Shed decks are simple, uncomplicated; and as Fair noted in several of his studies, the number of geometric design variables are few. These trays have high capacity/fouling resistance; yet, low efficiency. But should engineers think of sheds in a different way?

While Distillation practitioners seek improvements to the cross-flow or counter-flow tray designs, they rarely evaluate technology that simplifies the designs. The same practitioners consider sheds a necessary efficiency downgrade to gain fouling resistance. However, is this a fair evaluation for shed decks? A counter-flow tray revamp of a cross-flow trayed tower is rarely a one-for-one basis; so why do we apply the same limits to a potential shed deck revamp of a cross-flow tray. The authors propose to deviate from the traditional evaluation protocol and to consider a non-one-for-one scenario when revamping a conventional tray with a shed deck in a fouling system.

The authors will share a case study of several recent applications where shed decks replaced sieve trays on a non-one-for-one basis without a loss in operating efficiency. In each application, the shed deck conversion was preceded by multiple revamp iterations with designs intended to improve anti-fouling characteristics. The previous revamps not only provided limited success in improving run-length, but also provided lower hydraulic capacity than the final shed deck revamp.

549959 Using Dividing Wall Columns for Gasoline RVP Control

Using Dividing Wall Columns for Gasoline RVP Control

By David Kockler, GTC Technology

Reformulated Gasoline (RFG) standards that were imposed by the Environmental Protection Agency during the 1990s restrict the RVP of gasoline in order to reduce VOC emissions. Petroleum refiners struggle to meet RVP limits required by current RFG standards, especially during the summer season when RVP limits are the most stringent. One major cause for this problem is that distillation columns that were designed prior to the implementation of RFG standards generally do not provide sharp separations between C4 intermediate products and C5+ gasoline blend stock. A revamp of existing distillation columns to Dividing Wall Columns (DWCs) can be a very useful tool to help refiners control gasoline RVP.

GTC has several DWC projects that will be cited as examples during the presentation.

dkockler@gtctech.com

GTC Technology
900 Threadneedle St., Suite 800

Houston, TX 77079

281-597-4800

551701 Revamp to Two Parallel Dividing Wall Columns

Commercial implementation of Dividing Wall Columns (DWC) has been widely successful in the last two decades owing to robust simulation modelling and technology available for the design and operation of these columns. In most cases, the applications have been limited to individual columns with flexible product parameters. This paper discusses the revamp of two columns with DWC technology at the largest refinery in India.

The original design consisted of a naphtha splitter followed by a depentanizer column. Light Naphtha was separated to a mix of C5 cut and C6+ naphtha. The client now had a requirement to produce three products from the light naphtha. These cuts were iC5, C5s followed by C6+ naphtha. Additionally, a spare Reformate Splitter had been identified by the client for a possible revamp to meet the new product specifications.

The paper explores the multiple design scenarios to meet the new design objectives. These include the following:

- Revamp to two conventional columns in an indirect sequence
- Revamp to two parallel side-cut columns
- Revamp to two parallel dividing wall columns.

The pros and cons of the different revamp scenarios are discussed. The paper also illustrates the challenges encountered in this DWC revamp along with the methodology behind the control scheme for the application.

Distillation Control: Structured PIDs or MPC?

Sigifredo Nino, P.Eng.

Summa Control Solutions Inc.

Montreal, QC, CANADA

SNino@SummaControlSolutions.com

Presented at the 2019 Distillation Symposium, AIChE Spring Meeting, New Orleans, Louisiana, March 31 – April 4, 2019

Distillation Control: Structured PIDs or MPC?

Sigifredo Nino, P. Eng. (SNino@SummaControlSolutions.com)

Process Control Expert

Summa Control Solutions, Montreal, QC, Canada

Abstract

Material Balance Control and Model Predictive Control, the way it should be. A Deisobutanizer (DIB) case study.

The main purpose of this paper is to give an educated answer to the question that frequently rises when implementing a multivariable control strategy: What should be used? Structured Distributed PIDs OR Model Predictive Control (MPC)?

This answer herein given is supported by the presentation of an application to Deisobutanizer (DIB). It will describe how Material Balance Control was used to guarantee the stability, performance and robustness required to have the consistent steady state stability that is prerequisite to the implementation of any supervisory optimizing application such as Model Predictive Control (MPC). The paper will conclude by recommending that the answer to the question should be PIDs AND MPC and asserting that these are not competing but complementary control strategies.

Structured Advanced Regulatory Control (ARC)

This is the third in a series of papers presented by the author showing case studies of successful application of **material-balance control to distillation**, proposed by Shinskey in 1964, as the **manipulation of one or more product streams for control of product composition**. In 1984 Shinskey formalized his proposition to addressing a methodic way of designing multivariable control strategies using distillation control as the embodiment of a practical industrial multivariable application. The author, with Greg Shinskey himself on his corner, has applied and improved his proposal over the past 13 years of working with it, Nino (2011, 2017, 2018).

This paper describes an application that involves the splitting of two isomers of hydrocarbons, butane and isobutane, that having identical molecular weight, and boiling points that are very close to each other, therefore requiring a large number of plates and reflux ratios much higher than one (RR>>1). A good fit to material balance control. This column was designed to replace an older one, whose initial analysis had shown substantial economic benefits from increasing boilup to improve isobutane recovery but that could not be handled by the old tower due to the constraints imposed by the physical equipment condition.

As previously narrated by the author, the methodology starts by assigning the different species to four components, hence having the all the data required from the engineering design application software, the light key (LK) is isobutane (iC4), the heavy key (HK) butane (nC4), the heavier than heavy and lighter than light assigned to the balance. The model calculated a relative volatility of 1,272 and a D/F of 0.55 (See Table 1 for the details)

KEY		Feed	Distillate	Bottoms	
		z%	y%	x%	
Lighter	LLK	0.800	1.454		
Light	LK	55.000	84.996	18.330	
Heavy	HK	43.500	13.550	80.114	
Heavier	HHK	0.700		1.556	
Theoretical Stages			50	D/F =	0.550
Reflux Ratio (RR)			4.3	α =	1.272

Table 1- Isobutane splitter

The next step of the methodology should have been to use Table 1 information to plot the operating curves in order to assess the control-loop interactions. However, the model used to perform the calculations and perform the plotting was no suitable for this tower conditions. Notwithstanding all the flawless previous performance of Shinskey's methodology supported by the Douglas et al. model (1979), this event serves as a reminder that models, whether they are empirical or analytical, are reduced representations of the reality and will be a matter of time before a set of conditions will not fit in.

Nonetheless, the relative gain analysis being so vital in the methodology and with the commercial column simulator available, it was used to iterate the D/F ratio and to generate the slopes around the products composition specifications. Subsequently, this allowed the calculation of the relative gains by using the equation (Eq. 1)

$$\lambda_y(u_1, u_2) = \left[1 - \frac{\partial y/\partial x|_{u_1(y)}}{\partial y/\partial x|_{u_2(x)}}\right]^{-1} \qquad \text{(Eq. 1)}$$

The gradient in the numerator is related to the changes of a component in the top and bottom products, while keeping the top composition manipulated variable u_1 constant, and the denominator the corresponding ratio when the manipulated variable chosen to control the bottom composition u_2 is kept constant.

On a side note, Harry Kooijman (Shell Global Solutions) and Ross Taylor (Clarkson University) have developed ChemSep ®, a column simulator for distillation, absorption, and extraction operations. The author used the ChemSep 7.41 Lite and was able to successfully cross check the results given by iterating the D/F ratio in some of different column models readily available in the application and the calculations given by Shinskey's methodology that uses an empirical model. ChemSep is a very user-friendly, the Lite version is freeware, no strings attached, no endorsement either, just an unbiased comment from the author.

Back in the topic at hand. Using the column multicomponent separation process simulator, the calculations were done as follows: The slope for D or B flow constant is equal to -B/D, and iteration of D is required in all the cases. For example, constant boilup, V=L+D, is found by increasing L by the same amount D is decreased, and constant L requires the reflux to remain constant when D is changed.

Since the most of the benefits would come from manipulating the boilup to maintain the bottom control in check, having established the variable that would be best to control top composition, only the second row of the Relative Gain table was calculated (Table 2). The most favourable pairing would be the SV pairing with 1.01.

Relative Gains		**MV$_2$→y**		
		D [D/F]	L [L/F]	S [L/D]
MV1↓ x		-0.800	0.444	0.007
B [B/F]		n/a		
V [V/F]	0.652	0.449	3.142	1.010
R [V/B]				

Table 2 – DIB Relative gains for the pairings of interest

Choosing the right strategy comes from the objectives the customer has for the distillation tower, and from the practical limits that may rule out some of the options; markedly, the use of fired heater reboiler to control bottoms level. The most important outcome from the RG analysis is the selection of the best pairing in a dual composition control, a scheme that is known to minimize the energy consumption when both products are held at their specifications. Yet, another option in the selecting the right strategy for a distillation column derives from the tangible economic incentives associated to optimizing the recovery of the most valuable product. In this case associated to the maximization of the recovery of iC4 given the large value difference with the nC4, and the need to minimize the isobutane makeup to satisfy the total Alkylation plant demand.

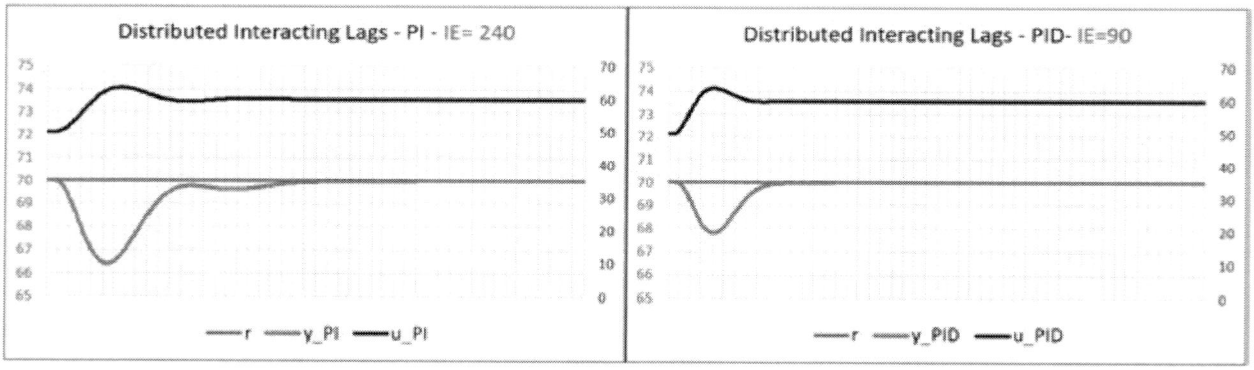

Figure 1- Distributed Interacting Lags - PI and PID application

The strategy proposed and implemented is sketched in Figure 1. Testing showed that the top control tray temperature had a strong and useful relationship with the impurity, nC4, contents in the distillate product, so it was used as a surrogate of the composition in a cascade scheme. Although never made explicit in any of the previous papers, the surrogate temperature controllers used in these applications are PID and not a PI only; and what is the difference? The reader may ask. Well, it is frequently found in the papers written to show the superiority of other control algorithms over the PID that the controllers are only PI. Distillation Column Composition follows interacting distributed lags dynamics that are much more responsive to the derivative action, and so the addition of the third mode to the PID algorithm can improve the control loop performance, measured as the Integral of the Error, by a factor of more than two, as can be seen in the plots of Figure 1. It should also be noted that the maximum deviation of the controlled variable from its setpoint has been noticeably reduced as well as the time to recover from load disturbances.

Knowing that the column pressure would be part of an optimization strategy, the tray-5 temperature was pressure-compensated by using the linear relationship proposed by Shinskey (1982):

$$T_{compensated} = T_{measured} - \left.\frac{\partial T}{\partial p}\right|_x (p_{measured} - p_{base}) \qquad \text{(Eq. 2)}$$

Where the partial derivative is the slope of the vapour-pressure curve for the desired mixture at the point of measurement, which can be estimated from either a test or from the Hadden and Grayson nomogram (Van Winkle, p. 106)

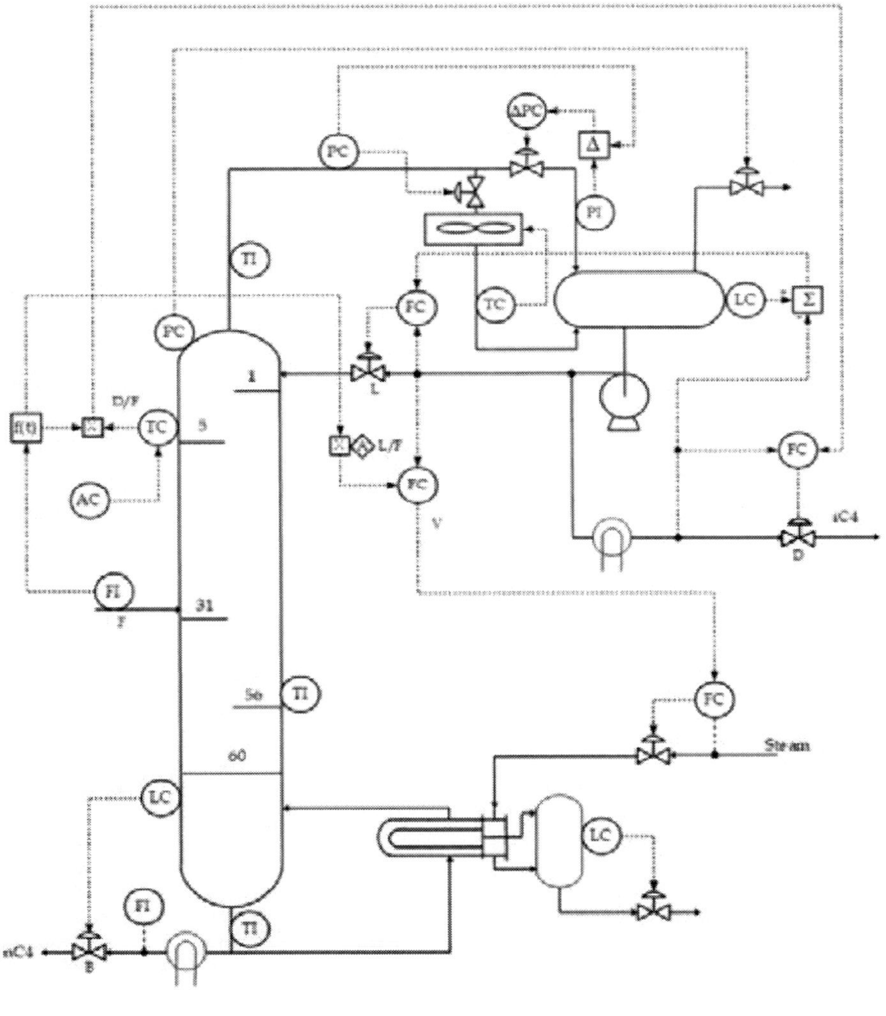

Figure 2- DIB Advanced Regulatory Control scheme

The factor that affects the most the product composition is the product split between top and bottom products, expressed by D/F. In consequence, in a column where only one product composition needs to be maintained, the external material balance needs to be precisely controlled. In this case, the biggest benefit would come from the maximization of the iC4 to the top product while maintaining the nC4 just below the maximum specifications. The isobutane recovered in the distillate product is equal to the amount of salable product generated per unit of that component in the feed, mathematically expressed as

$$R_{iC4} = \frac{D}{F \cdot z_{iC4}} \qquad \text{(Eq.3)}$$

Thus, the decision was to control the top temperature manipulating the D/F ratio along with a feedforward from Feed to set the Distillate product, and a reflux to feed ratio controlling the fractionation required to do so, by setting up the boilup. The overhead drum level is controlled by reflux and the Van Kampen decoupler guarantees a timely response of the D composition control, by eliminating the overhead imposed by the overhead accumulator lag.

In spite the existence of similar unworkable control strategies at the plant, the column pressure control system was originally designed and delivered to have a pressure differential controller across the hot bypass valve. Hence, being aware of a "quick-fix" solution proposed by Friedman (2008), the high-pressure measurement of the pressure differential controller was replaced by the setpoint of the tower pressure controller. The second modification to the original design was to change the energy input control strategy that used the condensing pot level controller to regulate the heat transfer area by changing channel head condensate level; its dynamic response is sluggish, nonlinear and does not allow the use of feedforward. The preferred strategy uses the steam admission valve to form a steam flow loop to control directly the boilup, and the condensing pot outlet valve is set to control the level at a setpoint slightly below the heat exchanger tubes bundle. This way the steam flow dynamic response is not only better but also linear; ideally the condensing pot level setpoint should be such that the tubes bundles is completely available for heat exchanging, as this prevents the corrosion at the steam-condensate interface, however it is expected that in a new installation some of the tubes may need to be covered by condensate to compensate for the reboiler oversizing.

In a previous application, the L/F ratio although available as a handle to optimize the ratio, it was never required to change from its commissioning value. In this tower it was evident from the beginning that this could be the handle for a supervisory optimizing application, as will be shown in the next section. Lastly, the material balance is closed by the base level controller, that manipulates the bottoms product flow.

Model Predictive Control (MPC)

None of the previous cases the author has presented (Nino, 2011, 2017, 2018) required any supervisory control strategy in place. In this tower it was possible to improve even further the optimization by pushing the operation unit against its constraints.

The optimizing application started by adding inferential models for the top and bottoms product impurities, as this will provide the ability to schedule the tower based on the minimization of the makeup iC4 DIB. The approach also opened up the possibility of closing a loop to maximize the L/F on line. MPC was chosen to implement the supervisory control application.

The two products impurities are estimated from inferences using General Distillation Shortcuts (GDS), methods (Friedman, 1985). The Top impurities, nC4, inference is used as the controlled variable and corrected by an analyzer reading, and the bottom impurities, iC4, is not corrected by either analyzer nor a laboratory sample, as the purpose of this inference is to make sure that there are not more iC4 to recover from the tower bottoms, namely it is used as an indicator that there is no more recoverable iC4 in the bottom of the tower. The Figure 3 shows the measurements used in the inferential models; these two variables are part of the MPC matrix

T_{tray5}	
T_{OH}	$y_{nC4\ (infer.)}$
T_C	GDS
L	nC4 in D
P_{top}	
D	

T_{tray56}	$x_{iC4\ (infer.)}$
Q_{steam}	GDS
T_B	iC4 in B
B	

Table 3- DIB tower - Impurities Inferential Models

The MPC overall control strategy objective is to maximize the recovery of the iC4 in the Distillate product (D), as this will decrease of make-up isobutane required to run the alkylation unit. When the makeup iC4 is not required, then the content of nC4 becomes important and must be maintained. There are economic benefits, related to energy savings, associated to the tower pressure minimization, subject to the limitations imposed by the cooling capacity, and the iC4 recovery is maximized by manipulating the L/F ratio, a handle that was readily available from the ARC implementation. The Table 4 displays the model predictive control matrix, the asterisk indicates the pairing for which a model was found, abbreviations and variable definitions can be found in Table 5. The different MPC technologies are quite similar and their principle of operation can nowadays be found in multiple textbooks, therefore neither time nor space will be spent in this article to address it.

Succinctly, the MPC objectives are:

- To maximize Top control tray temperature, so the iC4 make-up is minimized
- Whenever the make-up is not in service, to keep the nC4 impurities under the specifications
- To minimize the tower pressure, subject to the available cooling constraint
- To maximize L/F to improve fractionation
- To minimize the iC4 in the nC4, so the butane addition to the gasoline will not cause high RVP (Reid Vapor Pressure) values.

		MV			FFWD	
		$PC_{top}.SP$	$TC_{tr5}.SP$	L/F	F	T_C
	$y_{nC4\ (infer.)}$		*			
	L			*	*	
	PV_{top}	*	*	*	*	
C	LV_{bottom}	*	*		*	
V	ΔP_{twr}			*	*	
	P_{stm}		*	*	*	
	ΔT_C	*				*
	F_{iC4}		*	*	*	
	$x_{iC4\ (infer.)}$		*			

Table 4- MPC control matrix

Conclusion

Structured Advanced Regulatory Control, Inferential Composition Models and MPC have been successfully combined to provide the customer substantial economic benefit, in the 7 figures range. Doing the math, it will give an ROI in the order of days; however, it must be kept in mind that this is going to be a sustained revenue stream for the life-cycle of the unit operation! The numbers will then get into AWE-land.

Weighing on the PID vs. MPC Affair

The author has been involved in several projects where successful advanced regulatory controls have been the enablers of successful model predictive control applications. It takes a team of professionals who understand the benefits and limitations of both technologies to deliver real sustainable value to a chemical processing facility. We believe we have arrived to a common ground where it is understood that the technologies complement each other, sometimes ARC by itself is enough, sometimes ill-conceived MPC applications have been replaced by successful ARC, and very often the regulatory and advanced regulatory controls have been prepared so an effective MPC application can deliver the benefits expected by the customers. This is the way it should be.

For the sake of fairness, the author felt compelled to bring this subject in: Most of the published technical literature keeps focusing on how much better is the MPC compared to distributed PIDs; as far as Distillation Control is concerned, the oddest part is that almost all the PID-based structures used in the articles are either the conventional energy strategy, LV, that is very ill-suited for high reflux ratio cases, so the controllers must be severely detuned, or they use the material balance DV, which without the Van Kampen Decoupler is very slow responding. On top of everything, PIs instead of PIDs controllers are used. As mentioned above, the addition of the derivative mode to a controller will allow the reduction of the Proportional Band and Integral Time to the extent that a reduction in the integrated error by a factor of two will be attainable.

Indeed, the addition of Derivative mode will not overcome a deficient control strategy. It has to be part of a properly structured advanced regulatory control strategy.

Not to extend unnecessarily culmination of the paper, it has to be mentioned that there have also been editorial notes describing limitations, issues, and failed MPC applications. The author's goal with this paper is to bring this "needless battle" to rest, as it doesn't bring anything but confusion to the already troubled world of process control. What should be delivered to the customer is profitable, easy to understand and easy to maintain application; there is no magic or control algorithms that will violate the laws that govern the plant floor unit operations. Understanding of chemical engineering and process control fundamentals aligned with common sense should prevail. Let us finish with a quote the author found in Rossiter's MPC book (2018): "You need a good reason not to use a PID type of approach"

B	Bottoms product
D	Distillate product
F	Tower Feed
F_{iC4}	iC4 make-up flow
LV	Level valve
P_{stm}	Reboiler Pressure
P_{top}	Tower Top Pressure
PV	Pressure valve
Q_i	Heat Input
T_C	Fin-fans outlet Temp.

T_{OH}	Overhead vapor Temp.
T_X	Temperature at X
x_H	Heavy Key in B
x_L	Light Key in B
y_H	Heavy Key in D
y_L	Light Key in D
z_L	Light Key in F
ΔP_{twr}	Press. Diff. across Tower
ΔT_C	Temp. Diff. across Condenser

Table 5- Variable definition and abbreviations

On a side note, the author has been perusing through some of articles on Divided Wall Column control (e.g. Donahue et al, 2016), to realize that the same premises found for binary and sidestream distillation application holds. On the up side, the comparison of ARC and other technologies, such as MPC, has been more composed.

Bibliography

Donahue M. M. et al (2016) "Dividing wall column control: Common practices and key findings", Chem. Eng. And Proc., V.107, p.106-115

Douglas J. M.; A. Jafarey and R. Seemann (1979) "Short-Cut Techniques for Distillation Column Design and Control. 2. Column Operability and Control", Ind. Eng. Chem. Process Des. Dev., V.18, No.2

Friedman, Y. Z. (1995) "Simulation Based Inferential Controls", AIChE Spring Conference, Houston (TX)

Friedman, Y. Z. (2008) "Condenser Hot Vapor Bypass Control", Hydrocarbon Processing, June

Kister H. Z. (1990) Distillation Operation, McGraw-Hill, Boston (MA)

Kooijman, H. A.; R. Taylor (2019) The ChemSep Book, http://www.chemsep.com/book/ public domain.

Lieberman N. P.; E. T. Lieberman (2003) *Working Guide to Process Equipment*, McGraw-Hill, New York, NY

Luyben, W. L. (1975) "Steady-state Energy Conservation Aspects of Distillation Column Control System Design", Ind. Eng. Chem. Fundam., V.14, No.4

MacMullan E. G., F. G. Shinskey F. G. (1964) "Feedforward Analog Computer Control of a Superfractionator", Control Engineering, March

Nino, S. (2019) "*Distillation Control*" Section in Chapter 8 of *Process / Industrial Instruments and Controls Handbook, 6th Edition*, McMillan G./Vegas H. (Authors), McGraw-Hill, New York

Nino, S. (2018) "Distillation Control – Practical Aspects", AIChE Spring Conference, Orlando (FL), April 22-26

Nino, S. (2017) "Distillation Control: Still and Issue", AIChE Spring Conference, San Antonio (TX), March 26-30

Nino, S. (2011) "Improve Material Balance in High-Purity Distillation Control", Hydrocarbon Processing, November

Rossiter J. A. (2018) A First Course in Predictive Control, Second Edition, CRC Press, Boca Raton

Shinskey G. S. (1984) "From process knowledge to computer control: A methodology", Pacific Area Chem. Eng. Congress, Seoul (Korea), May

Shinskey G. S. (1982) *Distillation Control, Second Edition*, McGraw-Hill, New York (NY)

Shinskey G. S. (1996) *Process Control Systems, Fourth Edition*, McGraw-Hill, New York (NY)

Shinskey G. S. (1967) *Process Control Systems, First Edition*, McGraw-Hill, New York (NY)

Van Winkle M. (1967) *Distillation*, McGraw-Hill, New York (NY)

551705 Rolf Prince and His Work, Particularly Research into Distillation Technology

This work will discuss the outstanding research contributions of Emeritus Professor Rolf Prince AO, who passed away in Sydney aged 88, his remarkable personal story as a refugee saved from Nazi Germany and his outstanding professional career, recognised globally for his contribution to chemical engineering. Awarded the Order of Australia in 1997, he kept working at the University of Sydney for the next twenty years, stopped by illness only a few months before his death. Many have commented on his energy, and his frenetic pace only seemed to increase in his 'retirement'. His contributions to distillation technology, and particularly to sieve plate development and bio-distillation, will be reviewed in more detail in this paper.

551706 Distillation Technology: Is It Really Mature As We Approach 2020?

It has become fashionable in many circles to debate the status of an issue for the year 2020. Clearly many of these predictive efforts are valuable, but as the refrain goes "predictions are very difficult, especially when they concern the future". So, I thought I would talk about Distillation in 2020 as it stands in 2018, since really, changes in the next year and a half are probably already on their way.

Distillation is considered by many a "mature" technology. I don't know what that really means. Distillation is a mass transfer process, regulated by thermodynamics, and subject to the laws of conventional physics (both static and dynamic). We, as a whole understand the fundamentals but that does not make it a mature field. Descriptive methods of Distillation based exclusively on first principles and without empirical fits are not available still after several decades of research. We still rely on testing and empirical observations largely to develop and describe the field. We truly do not know what development will come around the corner. But, Distillation is such a prevalent process in Industry globally, that even very small improvements are capable of resulting in very large economic and sustainability benefits.

I asked someone one time why was it that the large research funding institutions did not fund Distillation Research. The answer was swift: "Because it is a Mature Technology". My next question was "How do you know it is a Mature Technology". Just as swiftly the response came: "Because no research is done in Distillation". This would be funny if it wasn't so prejudicial. Maturity of a Technology is not based on understanding the principles that rule it. It is based on how much progress can be made and what remains to be learned. Through this latter lens, Distillation is not mature and still has plenty of room for progress. Let us examine some of the progress the field has made in the last say 30 years in the following areas

Thermodynamics and Thermodynamic Efficiency
Process Configurations
Process Simulations
Process Control
Column Internals

Column Peripherals (Heat Transfer)
Reliability and Availability
Operation in Harsh Environments
Diagnostics

I will discuss each of these topics. Mention some of the great strides we have made in the field and elaborate on some of the strides we are still to make. Table 1 summarizes just some of the advances we have seen in distillation in the last 25 years. I have had the privilege to witness them all first hand. Also, I remember 25 years ago also hearing the comment about Distillation is a Mature Technology. Well, the table illustrates how a so called "Mature" technology has progressed and all these items were in the research phase in the last 25 years or so

551707 Debottlenecking Runtime of a Large Beer-Ethanol Column through a Fouling-Hydraulic Analysis and a Robust Equipment Design

The seventh largest bioethanol facility in the United States with a design capacity of 153 MGPY ethanol experienced major runtime problems. This state-of-the-art facility with the latest energy efficiency technology was commissioned in 2012. Years later, the plant pursued process improvements in the fermentation and distillation train. The ethanol yield was improved using new enzymes (bio-catalyst) under batch operation rather than continuous. This improvement resulted in the production of higher percentage of solids being fed to the distillation train. Thereafter, high-performance trays for the beer column were considered to increase runtime and to reduce energy consumption. Unfortunately, this upgrade impacted negatively the bioethanol production, the plant uptime decreased by eighty percent. This bottleneck lasted about 18-months, all this time the plant dealt periodically with the beer column cleaning-in-place (CIP) activities, and failed equipment modifications.

The plant seeking longer runtime contacted the original beer column equipment supplier. The supplier was updated with the operating issues and with the expectations for the replacement. Subsequently, a thorough fouling-hydraulic analysis of original and troubled high-performance trays was performed. Several scenarios were considered to explain the bottleneck impacting this 17-ft (5.2 m) diameter column. The complete analysis and column modifications were presented. These were accepted and swiftly and safely implemented. Conventional trays replaced existing high-performance trays. The new internals had features that assure uniform hydraulics to maximize mass transfer performance, and the most critical one, process uptime.

Since the debottlenecking activities in early 2Q2017, the bioethanol plant is operating at 168 MGPY and longer runtimes, and with expansion plans to produce 185 MGPY of ethanol.

This paper confirms that importance of a robust design of mass transfer equipment for critical applications, fouling in specific; and that in some applications, a *well-designed* conventional tray can "out-perform" a *poorly-designed* high-performance one.

549328 A Tray Column Design Method Incorporating a New Simple Efficiency Model

The concept of efficiency in distillation seems first to have been articulated in an article by W.K. Lewis published in Industrial and Engineering Chemistry in 1922. Since then that journal and its descendants have published many of the key papers on the topic of distillation tray efficiency (Murphree 1925, Lewis 1936, Chan & Fair 1984, Prado & Fair 1990, Garcia & Fair 2000). This paper starts with a very brief review of that history. This will bring us to a summary of our own recent work on efficiency in which we have first revised the famous and very simple O'Connell correlation to improve its accuracy as well as to broaden its applicability (AIChE Meeting, Spring 2017; CEP, July 2018).

Subsequently, Duss and Taylor (Distillation and Absorption 2018, Florence, September 2018) presented a new model for tray efficiency that uses the traditional concepts of numbers of transfer units for each phase but that is simpler than any analogous model in its demands for basic information and more accurate even than many more sophisticated models.

Any model based on the traditional concept of addition of resistances necessarily requires the slope of the equilibrium line. This, too, is a challenge for any models using the slope because it is the slope of a surface in hyperspace that is required and not so easy to estimate. Taylor and Duss (AIChE Meeting, Spring 2017) introduced the concept of Design Components as the basis for a new conceptually simple (but slightly computationally demanding) method of estimating the slope. In a more recent paper Taylor and Duss (Distillation and Absorption 2018, Florence, September 2018) provided an extremely simple equation to accurately approximate this slope that needs nothing more than the results typically immediately available from a column simulation.

What we attempt in this paper is to bring together these differing contributions to construct a complete and coherent tray design methodology that can take the designer from the early stages of a column simulation to preliminary sizing calculations all while using the least amount of information possible.

Missing from our earlier work is any discussion of physical property estimation. The simple efficiency model of Duss and Taylor requires only the gas and liquid phase viscosities. However, other popular models also require a diffusion coefficient. Diffusion in multicomponent systems is a complicated subject whose mathematical sophistication most often is ignored in column design work. This, more or less inevitably leads to the use of a pseudo-binary method of estimating diffusivity. Thus, we also compare and contrast some simple approaches to obtaining this quantity in a way that is as unambiguous as possible.

Our complete design procedure is illustrated with some practical examples.

551708 Results of an Alkylation Debutanizer Revamp with PFMD^{TM} Trays

High capacity trays have been available for many years. Distillation practitioners understand the unique intricacies of each offering; but many in the industry don't. When applying a high capacity tray device, it is important to understand both their strengths and weaknesses.

On small columns, applying high capacity trays required the practitioner to settle on which variable was more important for their system: capacity or efficiency. Many times to gain efficiency, more standard cross-flow trays were utilized as using a tray with more weir length would reduce efficiency.

The authors will share a case study of a recent application where a Debutanizer tower was revamped from traditional trays to high capacity trays. As the tower rectification section was not as liquid loaded as the stripping section, Parallel Flow MD trays were utilized. Revamp results show that PFMD trays transition the gap between a cross-flow and counter-flow efficiency; without a large loss in efficiency.

550565 Tray Design Challenges in Aromatics DWC

In 2011, ExxonMobil licensed a PXMax^{SM} paraxylene unit that employs a Dividing Wall Column to purify three products; Benzene, Toluene and mixed Xylenes. The tower internals are all trays in this case and there were some unique challenges to balance the pressure drop on both sides of the baffle. This paper will discuss the tray design employed by Sulzer and the challenges associated with that design. Tray design techniques for trayed Dividing Wall Columns will also be discussed. The tower has been in operation for a number of years and has been exhibiting better than expected tray efficiencies at design capacity.

551165 Measurements and Simulations of Natural Gas Separation Performance for a Pilot-Scale, High-Pressure, Cryogenic Scrub Column Fitted with Oldershaw Trays

A critical part of LNG processing is the removal of heavier hydrocarbons (Natural Gas Liquids or NGLs) from the gas stream prior to liquefaction. Understanding and accurately controlling this process is critical for optimal performance in terms of product quality, throughput, and reliable operation. LNG processing conditions offer up a different set of challenges compared to ambient and above ambient distillation. These challenges include models that do not adequately describe behaviour at extremes of temperature and pressure (typically <-60 °C and >40 bar), and a lack of empirical data describing the useful operating window for cryogenic hydrocarbon distillation. To better address these challenges, a unique pilot distillation facility [1] has been developed to replicate empirically the industrial-scale units. The random packing internals used previously were upgraded to tray internals as is usually employed industrially (27 Oldershaw trays were employed). The separation performance achieved was explored across a variety of operating conditions (temperature, pressure, inlet and outlet flows). Steady-state and dynamic models of the column were implemented within the process simulation software VMGSim. These models were validated with the experimental results and used to determine the column efficiency.

Results thus far indicate that under moderate conditions (-40 °C and 40 bar) there is little difference in performance between trays and random packing. Experiments with Oldershaw trays conducted at lower temperatures (-60 °C and 45 bar) resulted in a better separation performance when compared to experiments with trays run at higher temperatures. The design performance of the Oldershaw trays was confirmed by observations of liquid pooling as inferred from the response of the temperature sensors located above each tray. The results of varying feed flow rate to investigate its effect on tray efficiencies and determine the flood point will be presented and discussed. Future work, including operating at lower temperatures and higher pressures; as well as potential opportunities for collaboration will also be highlighted.

[1]. Hazewinkel, P., Sage, V., Khamphasith, M., Burke, N., May, E. F. *A New Pilot-scale facility for testing high pressure cryogenic distillation in LNG applications*. 2018 AIChE Spring meeting and 14th Global congress on process safety proceedings, Orlando, Florida, USA, April, 2018

552649 On the Fluid Dynamics of Inlet Downcomer Seal at Low Liquid Rates

The primary function of downcomers on trays is to facilitate countercurrent liquid flow down the column with vapor flowing up through the tray openings on sieve and valve trays. Downcomer liquid seal prevents vapor bypassing up through the downcomer thus enabling the downcomer to fulfil the proper function of stable liquid entry on to the tray deck. Under certain conditions, when downcomer seal is lost, vapor bypasses up through the downcomer, which inhibits the liquid passage through the downcomer. Downcomer seal loss is detrimental to overall column performance, particularly at low liquid rates under vacuum conditions.

The paper comprises of two parts. Part 1, courtesy of FRI, is video footage of downcomer seal loss and vapor blowing on a 1.22 m diameter sieve tray operating at very low liquid rates. Observations of vapor bypass up through the downcomer and the sieve holes were made with the aid of directional flow pointers. The second part focuses on initial checks for a tray designer to determine whether downcomer sealing has been realized. Those checks include downcomer clearance, outlet weir height, minimum weir loading, spray/blowing factor, downcomer backup, tray pressure drop and tray liquid holdup. A practical solution for mitigating downcomer unseal and extend a tray's operating range is by deployment of picket fence weirs.

543982 Test Validation of Sulzer Nexring 1.2 Random Packing

Sulzer NeXRingTM 1.2 random packings were recently testing at the Fractionation Research, Inc. commercial scale test facility. The packing was testing in both the xylene and butane test systems. Test data will be discussed along with relevant commercial applications.

551712 Fair Comparison of Packing Performance Data from Different Reference Systems

Engineers are responsible for insuring their company makes wise decisions when purchasing distillation column internals. This task is often complicated by a lack of "apples to apples" performance data on competing products - in which there are a common reference system, similar operating conditions, and equivalent-quality peripherals.

The authors have attempted to use available performance data and published correlations for mass-transfer and capacity for random and structured packing to create a fair comparison basis across different reference systems. This presentation will share our learnings to date.

549631 Simulating Maldistribution in Packed Columns

Maldistribution has long been known to be inevitable in packed columns and that it can significantly worsen the separation performance of a packed column. A recent presentation by L. Spiegel (The Maldistribution Story - An Industrial Perspective, Distillation and Absorption 2018, Florence) provides a convenient summary of the research on maldistribution up to relatively recent times.

Parallel column models sometimes are used to investigate the sensitivity of the HETP (the Height of an Equivalent Theoretical Plate) in a packed column due to maldistribution. In such models the vapor and liquid flows are split into two parts, which then enter one of two parallel columns. Billingham and Lockett (2002) proposed a two-column model for investigating maldistribution. Schultes (2000) studied maldistribution by modeling it with three parallel columns in a process simulator, and Stichlmair and Stemmer (1987) modeled maldistibution with no less than one hundred parallel columns! Although it is possible to develop a parallel column model using a process simulator, it takes considerable effort to arrange and specify the columns.

Recently, we have described what we call a Parallel Column Model PCM (Zhou et al., AIChE National Meeting, Orlando, 2018). While the model was developed with the aim of making it easier to model Dividing Wall Columns, we realized that it also is possible to use the ChemSep PCM to model maldistribution in packed columns. Both equilibrium stage and rate-based column models can easily be used within this framework. Our PCM can be used as a tool to study maldistribution inside any CAPE-OPEN compliant process simulation package.

In this paper we review the ChemSep PCM for DWCs and show how easily it may be used to describe maldistribution in packed columns. Cases considered examined by Fractionation Research Inc. (FRI), and summarized by Billingham and Lockett, as well as a case examined by Schultes, are used to study the liquid maldistribution with the help of our PCM.

References

Spiegel L., 2018, The maldistribution story - an industrial perspective, Chemical Engineering Transactions, 69, 715-720.

Billingham, J. F., and M. J. Lockett. "A simple method to assess the sensitivity of packed distillation columns to maldistribution." Chemical Engineering Research and Design 80.4 (2002): 373-382.

Schultes, M. "Influence of liquid redistributors on the mass-transfer efficiency of packed columns." Industrial & engineering chemistry research 39.5 (2000): 1381-1389.

Stichlmair, J., and A. Stemmer. "Influence of maldistribution on mass transfer in packed columns." Inst. Chem. Eng. Symp. Ser. Vol. 104. 1987.

Gaining Insights into Parting Box Hydraulics

By

Henry Z. Kister, Garry Jacobs
and
Abraham A. Kister
Fluor, Aliso Viejo, CA, USA

Presented at the Distillation Topical Conference, AIChE Spring Meeting, New Orleans, Louisiana, March 31-April 4, 2019

Gaining Insights into Parting Box Hydraulics

By Henry Kister, Garry Jacobs, and Abraham A. Kister, Fluor, Aliso Viejo, CA

Achieving good liquid distribution is crucial for the performance of packed towers and multi-pass tray towers. With the focus on liquid distribution, one device that has received surprisingly little attention in the published literature is the parting box. While good-practice guidelines are available, the published literature is short of work on the hydraulics of parting boxes and tested guidelines for its design.

The motive for the current tests was a hydrocarbon separation tower (designed by others, not Fluor) that was unable to achieve the design separation. A key suspect for the poor performance was the parting box whose design adhered to most good-practice guidelines. To gain insight into the parting box performance, we built a scaled-down model, geometrically similar to the parting box in the hydrocarbon tower, and tested it with water. This model was intended not only to provide insight into the performance, but also to provide generic insight into parting box hydraulics as tested in a realistic parting box design. The experiments conducted extended well beyond the specific design under investigation into generic parting box hydraulics. This paper describes our tests and findings.

Our tests showed the liquid in the test parting box was highly frothed up at both high and low liquid rates. The frothing was due to water jets from the feed pipe sparger entraining air into the water upon impact. This generated unexpectedly high froth heights in the parting box, and premature overflow at high rates. The liquid split to the five clusters of holes at the floor of the parting box was good at the maximum rates but quickly deteriorated as the liquid rate was turned down.

The use of a Perforated Impingement Plate (PIP) with either 20% or 10% open area, mounted either 2" or 4" above the parting box floor, greatly improved the split of liquid to the clusters of holes at the floor of the parting box. With a PIP, the split was very good to excellent both at the maximum rates and at turndown to about 50% of the maximum rates. Lower rates were not tested.

In the tests with PIP's, an air gap formed beneath the PIP, with tall froth above, producing premature overflows at high rates, and low and relatively less aerated froth beneath the PIP. The air gaps resulted from the water jets issuing from the feed pipe entraining air with them. This air descended through and accumulated beneath the PIP, pressuring up the space between the parting box floor and the PIP. The trapped air rose back relatively easily when the PIP had a large (20%) open area, incurring a relatively small pressure rise beneath the PIP and a relatively shorter froth backup above the PIP. The pressure buildup beneath the PIP and the froth backup above the PIP were intensified with the more restrictive 10% open area PIP.

The pressure buildup beneath the PIP helps maintain good liquid distribution at the outlet of the parting box. A troublesome observation is that the presence of the PIP generates a high froth buildup in the parting box, and at high rates, premature overflow, at well below the maximum expected capacity. In an operating tower, with vapor ascending around the parting box, such overflow may result in entrainment and maldistribution.

Our tests placed some of the existing parting box guidelines and practices under challenge. In particular, it showed that velocities as low as 5-6 ft/s of liquid exiting the feed pipe can cause intense frothing in the parting box that may be undesirable. The air gap phenomenon with PIP's, to our best knowledge first reported and explained here, adds a note of caution for PIP design. Premature overflows of froth above the top of the parting box have been observed due to frothing or due to air (or vapor) gaps below PIP's.

Parting Box Practice

It is well established that achieving good liquid distribution is crucial for the performance of packed towers and multi-pass tray towers. Kister's survey[1] attributed 80% of packed tower performance failures to maldistribution. Prominent among these are failures due to poor initial liquid distribution to the packed bed or multipass tray passes. The detrimental effects of poor liquid distribution were described by numerous excellent literature references (e.g., 2-5). Practical guidelines for distributor design are included in many texts and articles (e.g., 6-9).

With the focus on liquid distribution, one device that has received surprisingly little attention in the published literature has been the parting box. The parting box is the rectangular box that splits the incoming liquid to the troughs of a trough distributor. Figure 1 shows a common arrangement, in which an elevated parting box is mounted above the distributor troughs. A feed pipe, parallel to and elevated a short distance above the top of the box, and containing downward-pointing perforations (not shown in Figure 1), brings liquid to the box. The liquid may be overhead reflux or collected internal liquid mixed with liquid feed or pumparound return liquid. The parting box contains holes in its bottom, located above each of the troughs. These holes meter the liquid to each trough. The hole area above each trough varies, so that longer troughs, that supply the liquid to more distributor drip points, receive more liquid, while the shorter troughs with fewer distributor drip points receive less liquid. By design, a uniform liquid head is targeted in all the troughs.

A similar device is sometimes used to distribute liquid to multipass trays, except instead of feeding troughs, the parting box liquid is metered to each panel of the tray below to achieve the desired liquid split among the passes.

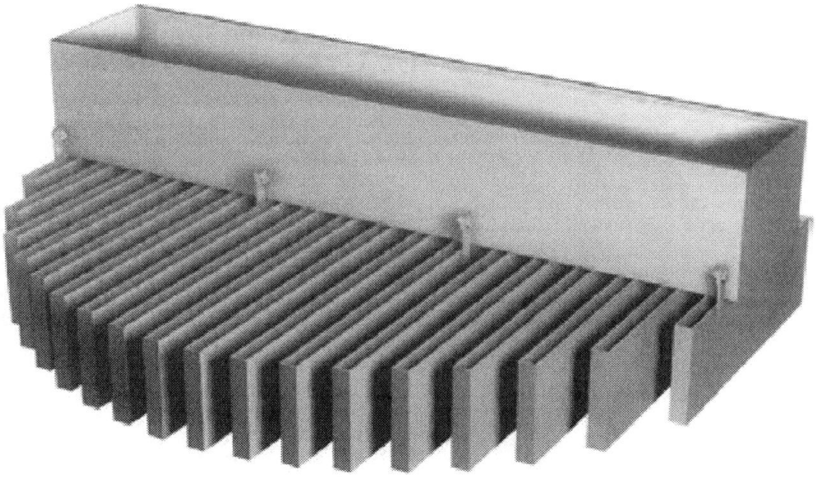

Figure 1. A typical parting box splitting liquid to the troughs on top of a packed tower distributor (Courtesy of Sulzer Chemtech)

Considering the importance of the parting box, the published literature is short of work on its hydraulics and guidelines for its design. Guidelines for parting box design were first described by the Norton Company[10] and later by Fadel[11]. These guidelines were endorsed, cited and supplemented in Kister's book[8]. Key recommended practices based on these sources include:

- Feed velocities leaving the pipe sparger should not exceed 10 ft/s[10] and preferably be less than 4-5 ft/s[11]. High velocities may disturb the liquid surface or cause excessive aeration in the parting box.
- Parting boxes should be sparger-fed[11]. The sparger holes must fall between the troughs and not over the perforations in the parting box[10,11].
- Sparger pipes should be oriented parallel to and centered directly above the centerline of the parting box[8].
- The sparger should be designed per good practices, such as those described in *Distillation Operation*[8].
- Sparger hole velocity should be considerably higher than the velocity in the distributor pipe. While uniform distribution out of the sparger perforations is not critical, as the parting box will redistribute the liquid, excessive horizontal velocities in parting boxes can generate waves and overflow at the ends of the boxes.
- Liquid fall height from the sparger into the parting box should be minimized to prevent excessive splashing, frothing, or aeration[8].
- Liquid depth in the parting box should not be too shallow, as surface disturbances in the parting box may cause uneven split to the troughs. Also, low liquid depth increases hydraulic resistance to flow along the length of the parting box, generating hydraulic gradients. Nor should the liquid depth be too high to approach an overflow.

Parting Box Tests

The motive to the current test series was a hydrocarbon separation tower (designed by others, not Fluor) that was unable to achieve the design separation. Fluor was assisting with the troubleshooting of the tower. A key suspect for the poor performance was the parting box design. Although most of the good practice guidelines above were adhered to, some were not, and others were marginal. The greatest concern was that some sparger holes were directly above the parting box perforations, and to a lesser degree, that the liquid velocities leaving the sparger holes were on the high side (6-7 ft/s). It was unknown to what extent these issues would affect the liquid distribution. It was then appreciated how little is available on the hydraulic performance of parting boxes.

To gain insight into the parting box performance, we built a scaled-down model, geometrically similar to the parting box in the hydrocarbon tower, and tested it with water. This model was intended not only to provide insight into the current issue, but also to provide generic insight into parting box hydraulics tested in a realistic parting box design. The experiments conducted extended well beyond the specific design under investigation into generic parting box hydraulics.

Test Rig and Procedure

Figure 2 shows a top view of the test rig. The parting box was rectangular acrylic 21.5" long x 2.75" wide x 9.5" tall, and contained a perforated floor plate 20.5" x 1.75" cemented to the bottom flange. The floor plate had 75 holes at 3/16" diameter drilled into it using a press drill, in five clusters as shown in Figure 2, mimicking the hole geometry and arrangement in the floor of the hydrocarbon tower parting box. The total perforated area was 5.8% of the floor plate. The top of the rig was saddle-shaped to support the 2.07" ID transparent PVC perforated pipe sparger. The sparger had 12 downward-pointing 3/8" holes, six equally-spaced on each side of the tee. The bottom of the pipe was 12.5" above the perforated floor plate. Hose clamps were used to fix the pipe along the parting box length so the center lines of the tee lined up directly above the center lines of the floor plate. Two 5/32" vent holes, each at the opposite end of the tee, were drilled at the top of the pipe. Loose hose clamps were slip-fitted around the pipe and moved over the vent holes to cover them after the air vented out at the beginning of each run.

The peripheral ½" wide flanges surrounding the hole field of the floor plate were used for mounting two peripheral ½" wide hollow rectangular inserts, one 4" tall, the other 2" tall, one above the other. A flat momentum-breaking perforated plate, referred to as Perforated Impingement Plate (PIP), such as that used in many commercial parting boxes to arrest the impact of the liquid jets issuing from the sparger on the liquid surface in the parting box, was sandwiched between the inserts. Two PIP's were used in the tests, one at a time, each containing 3/16" perforations, one with 10% open area, the other with 20% open area. The inserts permitted each PIP to be mounted either 2" or 4" above the floor plate.

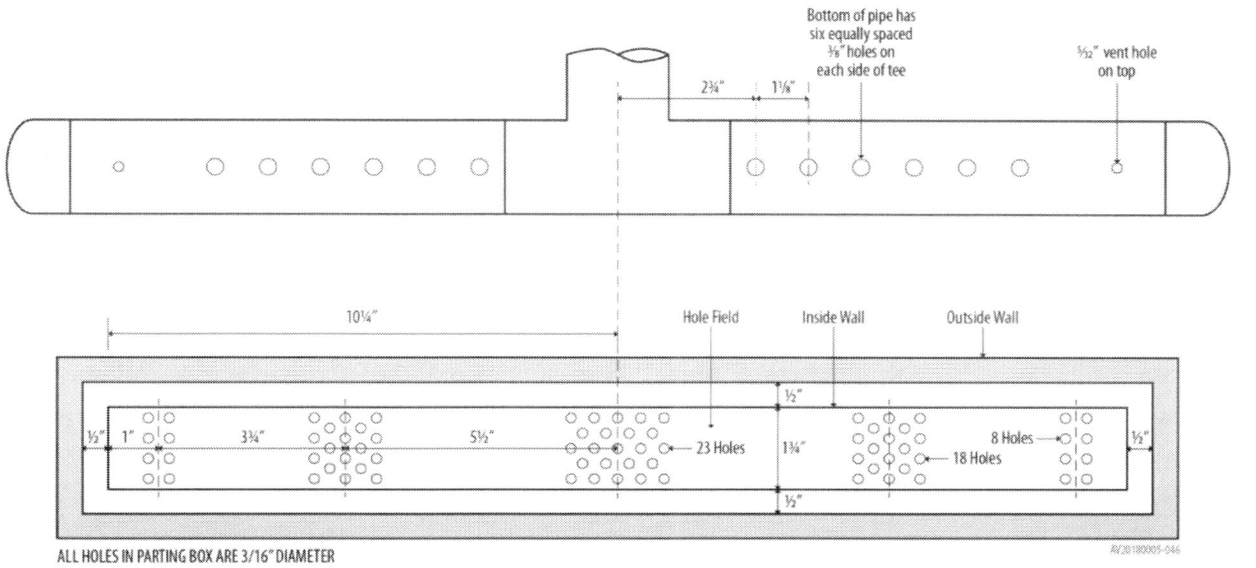

Figure 2. Test Rig Sketch: perforated pipe and parting box floor plans

Tap water to the rig was supplied by a 25 gpm submerged centrifugal pump from the bottom of a 96 gallon drum. A throttling valve in the pump discharge line to the sparger enabled manually adjusting the flow rate. For many runs, the valve was operated fully open and the pump delivered 25 gpm with a discharge pressure of 0.5 psi. For lower flow rates, the valve was set at a fixed throttle right at the start of the run. The water level on the drum was kept constant within plus or minus a foot by pumping water into it from two other drums, a 64 gallon drum and a 55 gallon drum, using two additional submerged pumps. In runs where more than 25 gpm were required, water to the rig was supplied by a 40 gpm centrifugal pump with valve throttling to the desired flow rate. The tests were conducted in the open air in dry weather with ambient temperatures ranging from about 65°F to 85°F.

Water leaving each cluster of holes in the floor plate descended into a separate chute, each diverting the collected water into a separate collector bin by gravity via a flexible pipe containing bellows. Each bin contained a ruler vertically fixed to the bin wall. At the end of each run a photograph was taken of the liquid level at the ruler in each bin. Prior to the tests, and again at the end of the tests, each bin was calibrated by stepwise introducing measured quantities of water and recording the water height in the bin. At the end of the tests the calibration was repeated, this time by stepwise removing the water from each bin by fixed quantities and again recording the heights in each bin after each step. The differences between the two calibrations were small in all the bins. Typically, in the larger bins collecting the liquid from the central three clusters, 1 inch of height corresponded to about 3 gallons, while in the smaller bins collecting liquid from the side clusters 1 inch of height coincided with about 2 gallons.

Each test was concluded after about 110-140 gallons were collected and the central bin approached its maximum desired height. Upon a signal, the pump was stopped and all the bellowed pipes were simultaneously turned upwards so the water remaining in the rig was trapped by the pipes and no longer entered the bins. After all the bin water height readings were completed, the water was pumped back to the feed drums for the next run. At the end of the tests, the water was pumped into garden irrigation hoses so it was all reused and none wasted.

Test Results: No Perforated Impingement Plate (PIP)

Table 1 presents key calculated hydraulic parameters for the feed pipe, its perforations, and the perforated floor plate of the parting box. For flow rates below 20 gpm, the sparger pipe hole velocities were within the desired velocity guidelines, and at 25-27 gpm slightly above the desired values but still within the recommended range. The pipe hole velocity was well above the pipe velocity, per the recommended practice guidelines. The calculated liquid head in the parting box was well below the overflow, as recommended. The only guideline not adhered to was that the two middle pipe perforations of each lateral fell directly above the second and fourth floor hole clusters. This pattern resulted from mimicking the actual tower under investigation.

Table 1. Calculated Pipe and Hole Velocities

Liquid Rate, GPM	27.00	25.00	20.00	12.00	Recommended guidelines (see earlier)
Pipe velocity, ft/s	1.29	1.20	0.96	0.57	Hole velocity>>pipe velocity
Pipe rho vee squared, lb/ft.s^2	104	89	57	20	
Total number of pipe perforations	12	12	12	12	
Pipe perforation diameter, in	0.375	0.375	0.375	0.375	
Pipe perforation velocity, ft/s	6.54	6.05	4.84	2.90	<10 ft/s, desired <4-5 ft/s
Pipe perforation rho vee squared, lb/ft.s^2	2667	2286	1463	527	
Number of floor perforations	75	75	75	75	
Diameter of floor perforation, in	0.1875	0.1875	0.1875	0.1875	
Floor perforation velocity, ft/s	4.18	3.87	3.10	1.86	
Floor perforation rho vee squared, lb/ft.s^2	1092	936	599	216	
Calculated liquid head in parting box, inches	6.53	5.59	3.58	1.29	Avoid overflow (<9.5")

Figure 3 shows photographs of the parting box at four different liquid rates with no PIP's. Figure 4 shows the corresponding measured liquid split among the various clusters.

Initially, there were no vent holes in the feed pipe. The upper photograph in Figure 3 shows a pocket of air that remained in the pipe, unable to exit. For the next run, we drilled two 5/32" holes in the top of the pipe to vent the air. Vent holes are common practice in commercial feed pipes. As can be seen in Figure 3, this eliminated the air pocket. From then on, air was vented via the vent holes at the beginning of each run. Once the air pocket was eliminated, we moved a hose clamp to cover each vent hole to prevent water jets from shooting up.

This experience provided us with an opportunity to evaluate the impact of the gas pocket in the feed pipe on the liquid split from the parting box. As the upper diagram in Figure 4 shows, the presence of the air pocket had practically no effect on the liquid split from the parting box at the 25 gpm flow rate.

We performed one more experiment at this rate. Initially, we installed the two inserts in the parting box with no PIP between them. This would permit close comparison with the runs that had the PIP. With this arrangement, the outermost perforations in the floor plate were about 1/8" from the inside wall of the inserts. For comparison, we re-ran the same test with the inserts removed altogether, so that the width of the active area was about 1" larger and the outer holes were about 5/8" from the inside wall. This test showed identical distribution and performance to the test with the inserts. All the tests that followed had the inserts, but in this series ("A") with no PIP's.

Figure 4 is a bar chart comparing the measured liquid split for these runs (various colors) with the expected split assuming every perforation in the floor plate delivers the same amount of liquid (blue bar). Figure 4 shows that the liquid split from the parting box was good only for the 27 gpm run (orange bar), where the froth height was approaching the overflow. The further the liquid rate was lowered, the more the liquid tended to issue via the middle cluster of holes, and the less through the off-center holes. The amount of liquid through the outermost holes did not deviate significantly from the expected in any of the runs.

Figure 3 shows intense frothing in the parting box at all the rates. The frothing occurred beneath the sparger pipe holes, while liquid directly under the unperforated tee remained clear. This liquid was very clear at the 12 and 20 gpm runs, but at the higher rates (25-27 gpm) froth built up above the clear liquid.

Figure 5 compares the froth heights as determined visually from the photographs to the liquid heads calculated from the orifice equation (with coefficients per Perry's Handbook[6]). The measured froth heights were higher than the calculated liquid heads, with the differences increasing with the liquid rate, reaching about 20-30% at the 25 - 27 gpm runs. The taller froth heights are due to the aeration in the parting box.

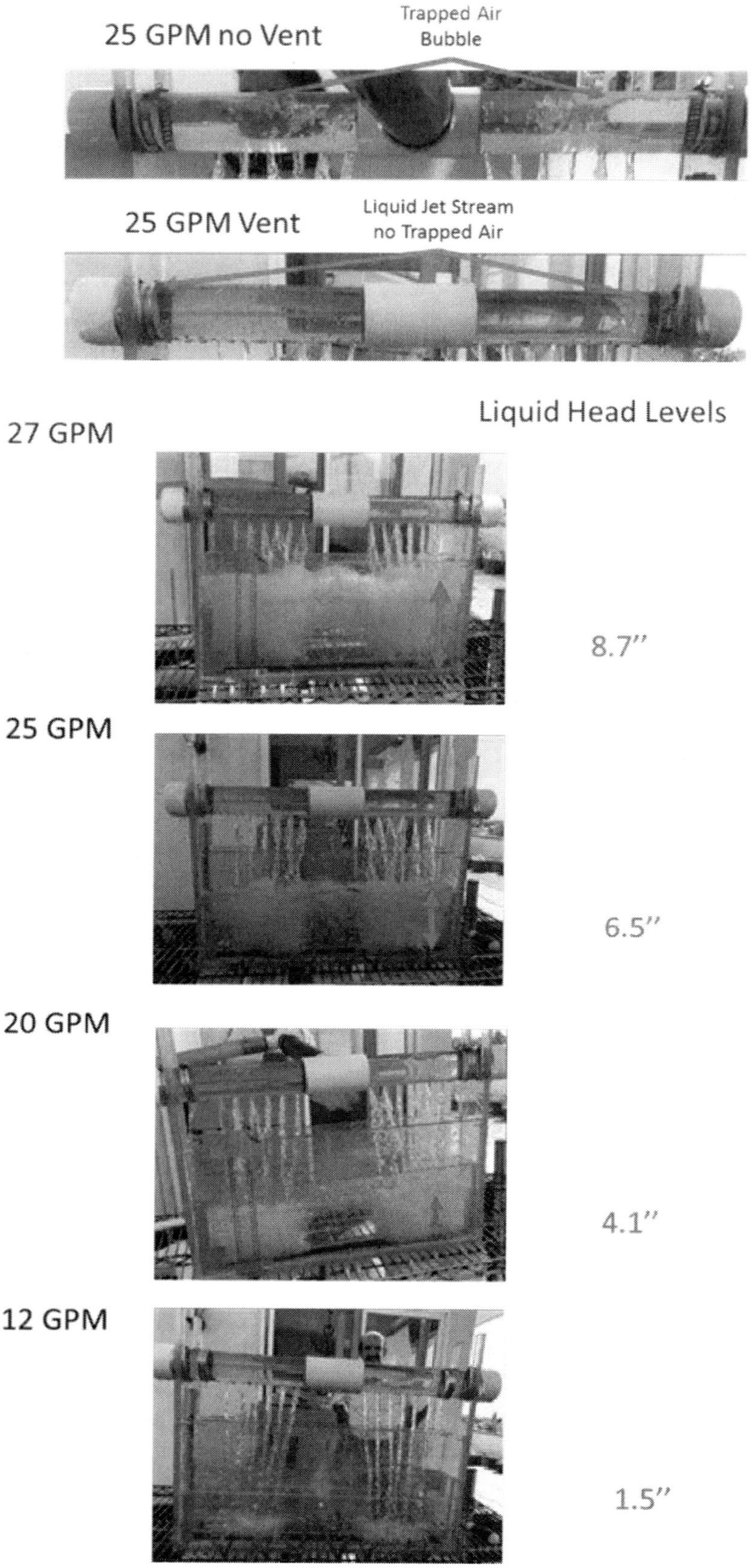

Figure 3. Tests with no PIP.

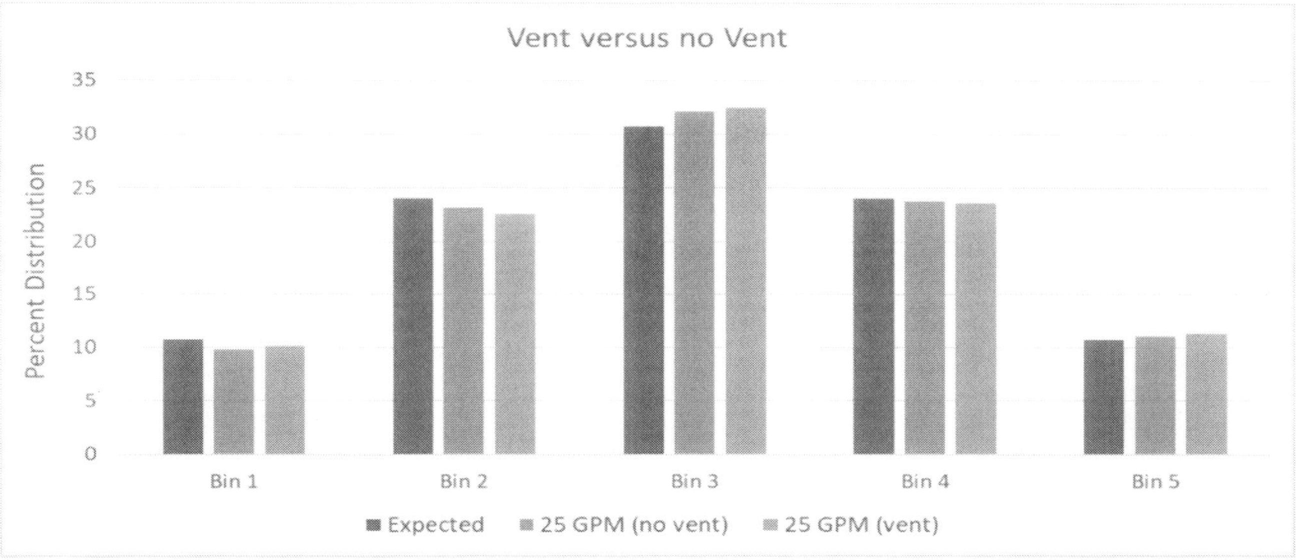

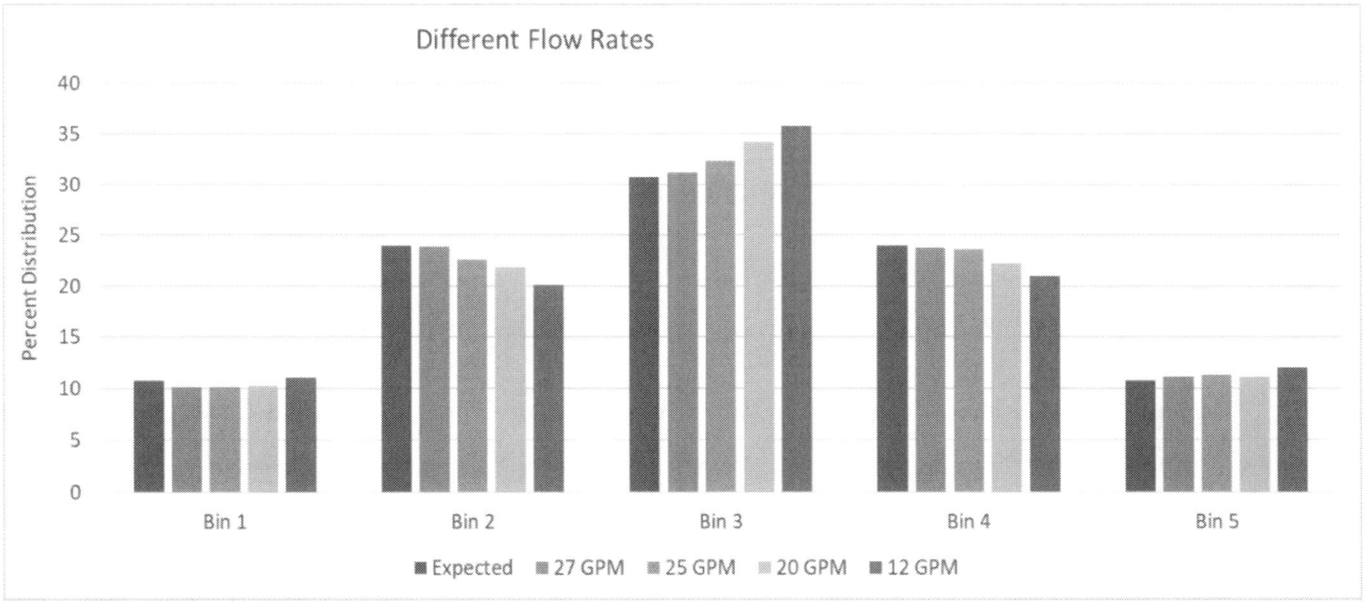

	Bin 1	Bin 2	Bin 3	Bin 4	Bin 5	Liquid Head (inches)
Expected	10.7	24.0	30.7	24.0	10.7	
27 GPM	10.1	23.9	31.2	23.8	11.1	8.7
25 GPM	10.1	22.6	32.4	23.6	11.3	6.5
20 GPM	10.2	21.9	34.3	22.4	11.1	4.1
12 GPM	11	20.1	35.8	21	12	1.5

Color	Percent Off from Expected
Green	0-0.99%
Yellow	1-1.99%
Red	2%+

Figure 4. Liquid split among the bins compared to the desired split based of equal liquid flow rate through each perforation, no PIP

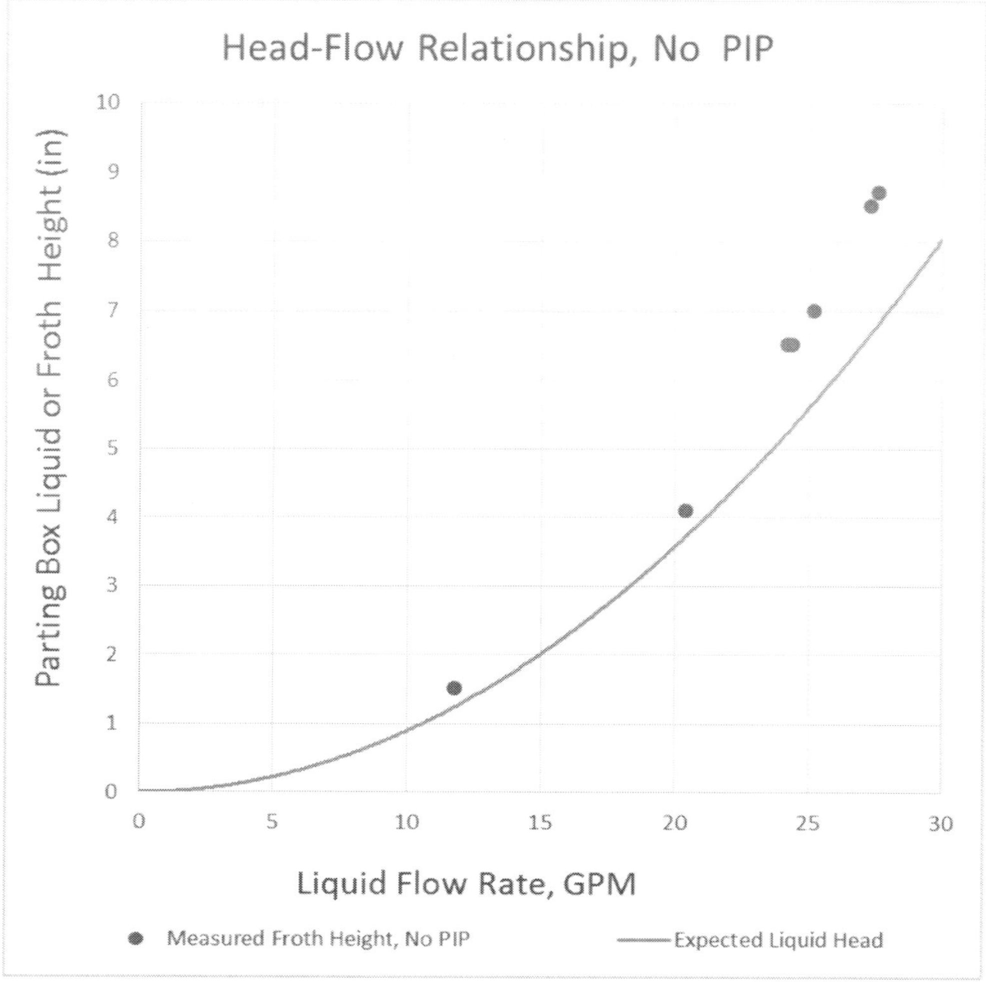

Figure 5. Measured froth heights in the parting box compared to the head-flow relationship, no PIP tests.

At the lower rates, the liquid beneath the sparger tee was quite clear, whereas, below the liquid jets, the liquid was somewhat aerated. At the higher rates, the aeration migrates into the liquid beneath the sparger, but remains less frothy than on either side. Aeration reduces static head (i.e., less clear liquid), resulting in liquid maldistribution that favors the less aerated central region (Bin 3), relative to the peripheral regions. As the aeration increases with rate, so too does the froth height and the deviation from expected clear liquid head in Figure 5. The practical implication is that overflow is likely to occur earlier than with non-frothy liquid.

One more observation from Figure 3 is that the liquid jets from the sparger pipe holes were quite even and pointed straight down, with only a slight horizontal slant in some of the runs. This is to be expected when the hole velocity is well above the pipe velocity as per the good practice guideline described earlier.

Test Results with 20% Open Area Perforated Impingement Plate (PIP)

The next series of tests inserted a 20% open area PIP at elevations 4" or 2" above the floor. It is important to keep in mind that, in comparison, the perforated area of the parting box floor plate is only 5.8% of the floor area. Figures 6 and 9 show photographs from the tests. Figures 7 and 10 show the measured liquid splits. At 27 gpm, an unexpected pattern was seen. There was a high froth buildup above the PIP, with a small overflow of froth out of the parting box, with a smaller layer (about 1.75") of much less aerated liquid on the parting box floor plate. This is clearly visible in Figure 6, but less so in Figure 9. A small reduction in liquid flow rate to 25 gpm transformed the froth to a pattern more similar to that without the PIP (Figure 3), but there was still a small air gap under the PIP. The turndown test at 12 gpm produced the expected pattern, with relatively lowly-aerated liquid at the parting box floor, almost to the height predicted by the orifice equation (Table 1).

Figure 7 and 10 show excellent split of the liquid at high rates. With the PIP 4" above the floor, the excellent split persisted in the 12 gpm (turndown) run. With the PIP 2" above the parting box floor and 12 gpm flow, there was a very slight indication of the previously-observed tendency for liquid to issue preferentially via the middle cluster of holes, and the less through the off-center holes, but the split was still very good and much better than without the PIP.

The good split at turndown with the 20% open PIP is emphasized in Figure 8 which compares the turndown runs with and without the PIP. The pattern produced with no PIP, in which the off-center bins receive less liquid, is still faintly visible, but the magnitude is much reduced.

Figures 9 and 10 show similar results but with the PIP mounted 2" above the parting box floor. Again, there was a large change in froth heights going from 27 to 25 gpm, and a slight overflow at 27 gpm.

Figure 11 compares the froth heights as determined visually from the photographs to the liquid heads calculated from the orifice equation (with coefficients per Perry's Handbook[6]) for the tests with the 20% open PIP. Two solid curves are shown on Figure 11. The orange curve is the same as in Figure 5, based on the floor orifices alone. The purple curve also includes the liquid head calculated across the PIP. Since the PIP and the floor open areas are 20% and 5.8%, respectively, of the parting box cross section area, the additional liquid head incurred by the PIP is small. The parting box froth height plotted in Figure 11 is the height above the parting box floor, not allowing for the PIP.

At the low liquid rate, the froth height in the parting box equaled the liquid head calculated from the orifice equation. At 26-27 gpm rates the froth height shot right up, becoming 30% higher than the liquid head calculated from the orifice equation (accounting for the PIP), suggesting a possible capacity limit flooding at the PIP. At these rates the froth in the parting box started to overflow. This was observed with the PIP mounted either 2" or 4" above the parting box floor.

27 GPM

> 5.5″ on PIP

1.75″ on bottom

25 GPM

2.2″ on PIP

3.25″ on bottom

12 GPM

0″ on PIP

1.25″ on bottom

Figure 6. Tests with 20% open area PIP, 4" above the parting box floor

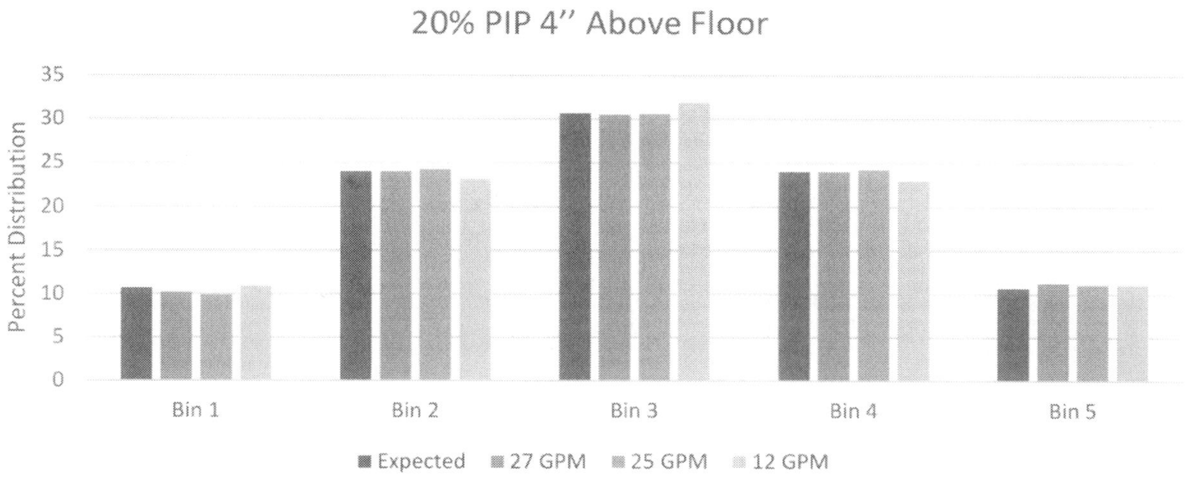

Figure 7. Liquid split among the bins compared to the desired split based of equal liquid flow rate through each perforation. 20% PIP, 4" above the parting box floor.

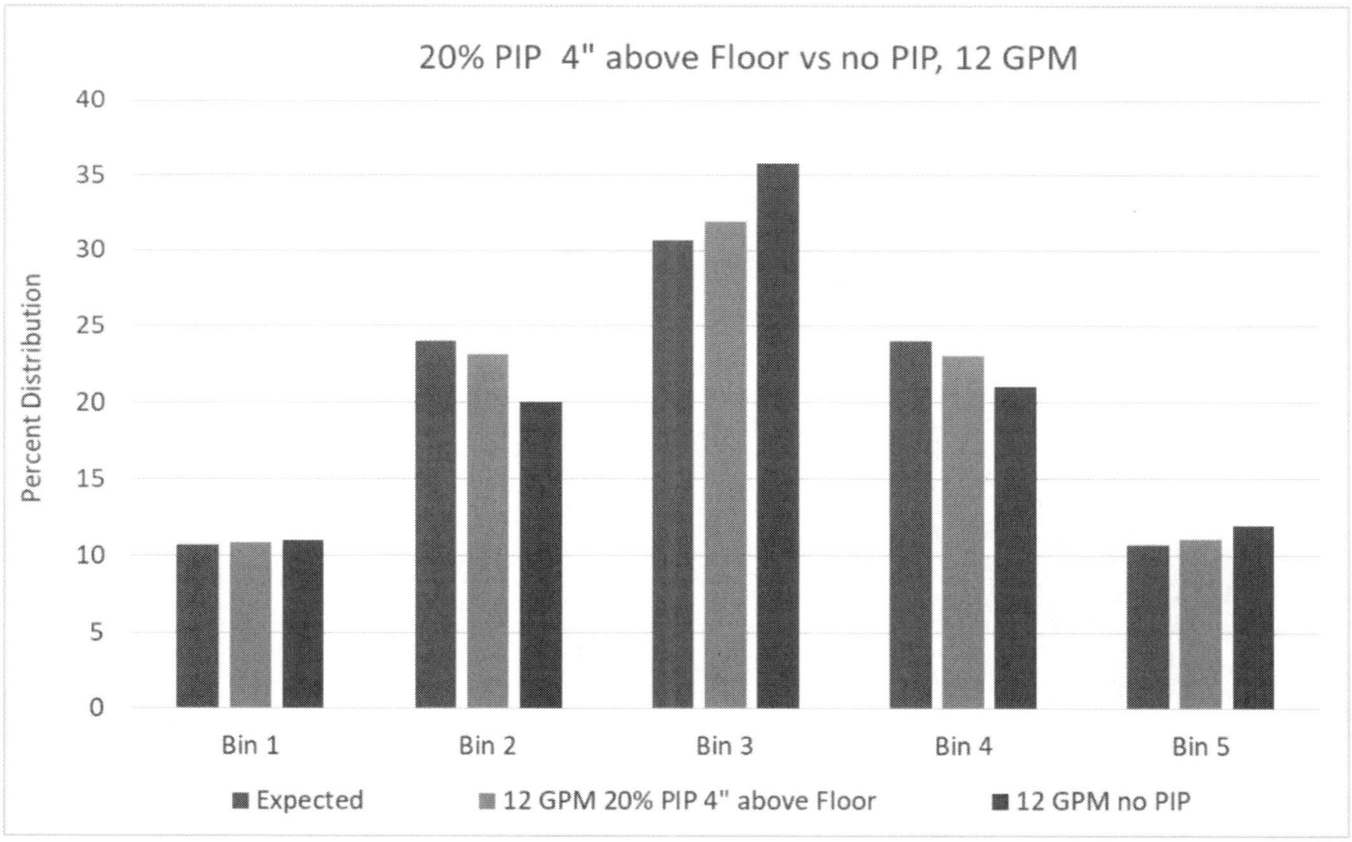

Figure 8. Liquid split among the bins compared to the desired split based of equal liquid flow rate through each perforation, at turndown rates, with versus without 20% PIP, 4" above the parting box floor

27 GPM

7.2'' on PIP

1.75'' on bottom

25 GPM

5.1'' on PIP

1.75'' on bottom

12 GPM

0'' on PIP

1.25'' on bottom

Figure 9. Tests with 20% open area PIP, 2" above the parting box floor

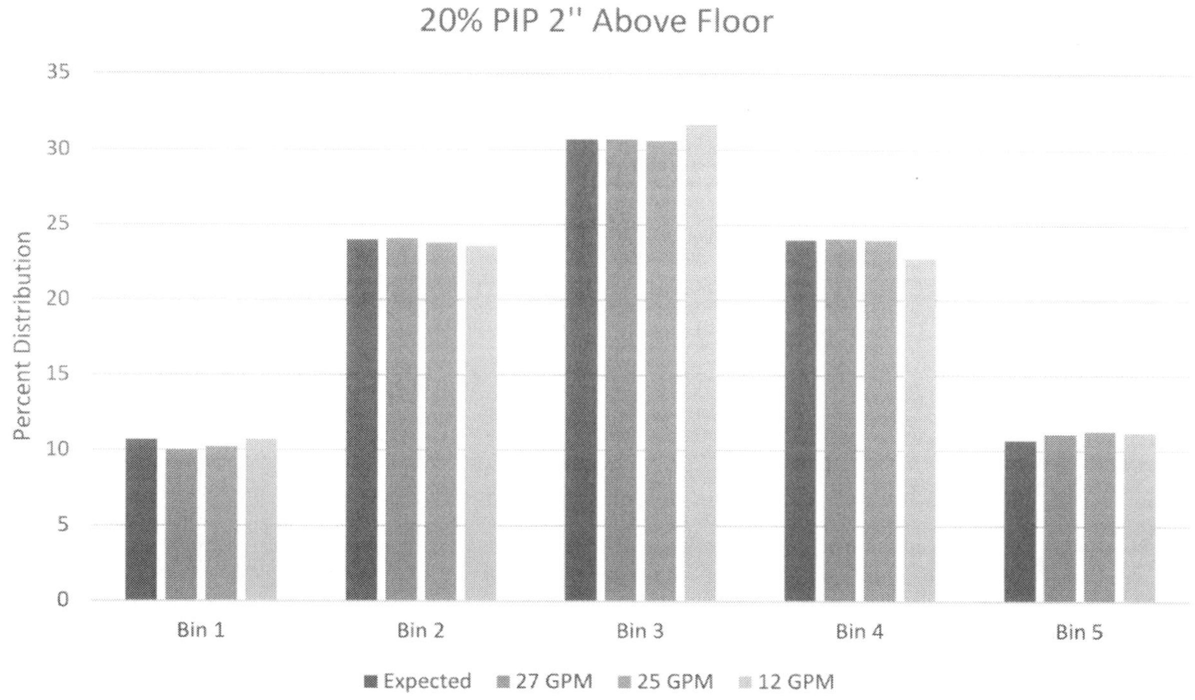

Figure 10. Liquid split among the bins compared to the desired split based of equal liquid flow rate through each perforation. 20% PIP, 2" above the parting box floor

Overall, the 20% PIP mounted either 2" or 4" above the floor provided an excellent liquid split to the holes, but also led to froth overflow at an earlier flow rate of 27 gpm. Without the PIP, there was no overflow at this liquid rate.

Test Results with 10% Open Area Perforated Impingement Plate (PIP)

The next series of tests inserted a 10% open area PIP at elevations 4" or 2" above the floor. It is important to keep in mind that in comparison, the perforated area of the parting box floor plate is only 5.8% of the floor area. Figures 12 and 14 show photographs from the tests. Figures 13 and 15 show the measured liquid splits.

The unexpected froth pattern observed at the 27 gpm tests with the 20% open area PIP intensified in the 10% open area PIP tests and was clearly visible in the 25 and even 12 gpm tests (Figures 12, 14). At 25 gpm, there was a high froth buildup above the PIP, with a small overflow, with a smaller layer (about 1-2") of much less aerated liquid on the parting box floor. This occurred at both elevations of the PIP (4" above the floor, Figure 12, and 2" above the floor, Figure 14). As liquid rates were reduced the same pattern persisted but the froth height above the PIP declined, reaching 4" and 1" at 20 gpm and 12 gpm, respectively.

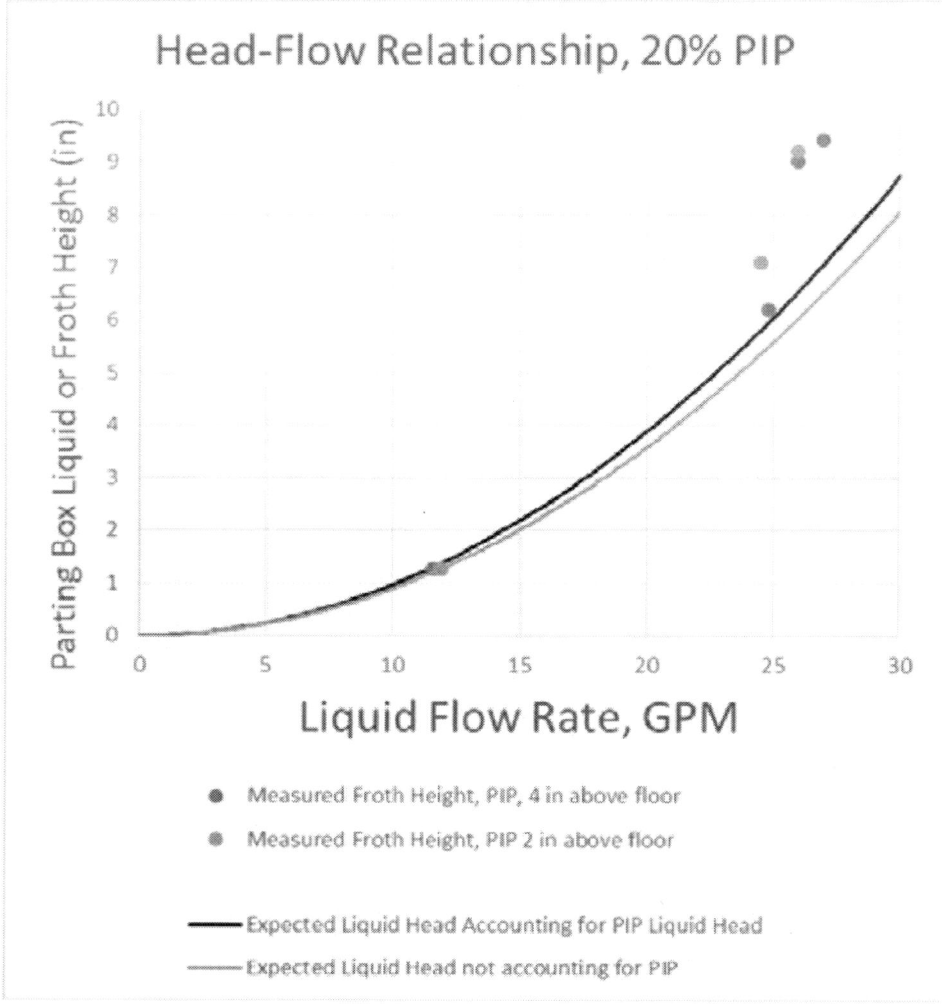

Figure 11. Measured froth heights in the parting box compared to the head- flow relationship, 20% PIP tests.

Above the PIP, the frothing occurred beneath the sparger pipe holes, while liquid directly under the unperforated tee remained clear. This is similar to the observation made in the no-PIP tests (Figure 3) but here the frothing and the clear liquid zones were above the PIP. This liquid beneath the tee above the PIP was very clear at the 20 gpm runs, but at the higher rates (25 gpm) a small layer of froth built up above the clear liquid.

In all these runs there was a 1-2" layer of less aerated liquid on the parting box floor. Figures 13 and 15 show very good split of liquid at both high rates and turndown to the various bins with the 10% open area PIP. The pattern produced with no PIP, in which the off-center bins receive less liquid, is still faintly visible at turndown, but the magnitude is much reduced.

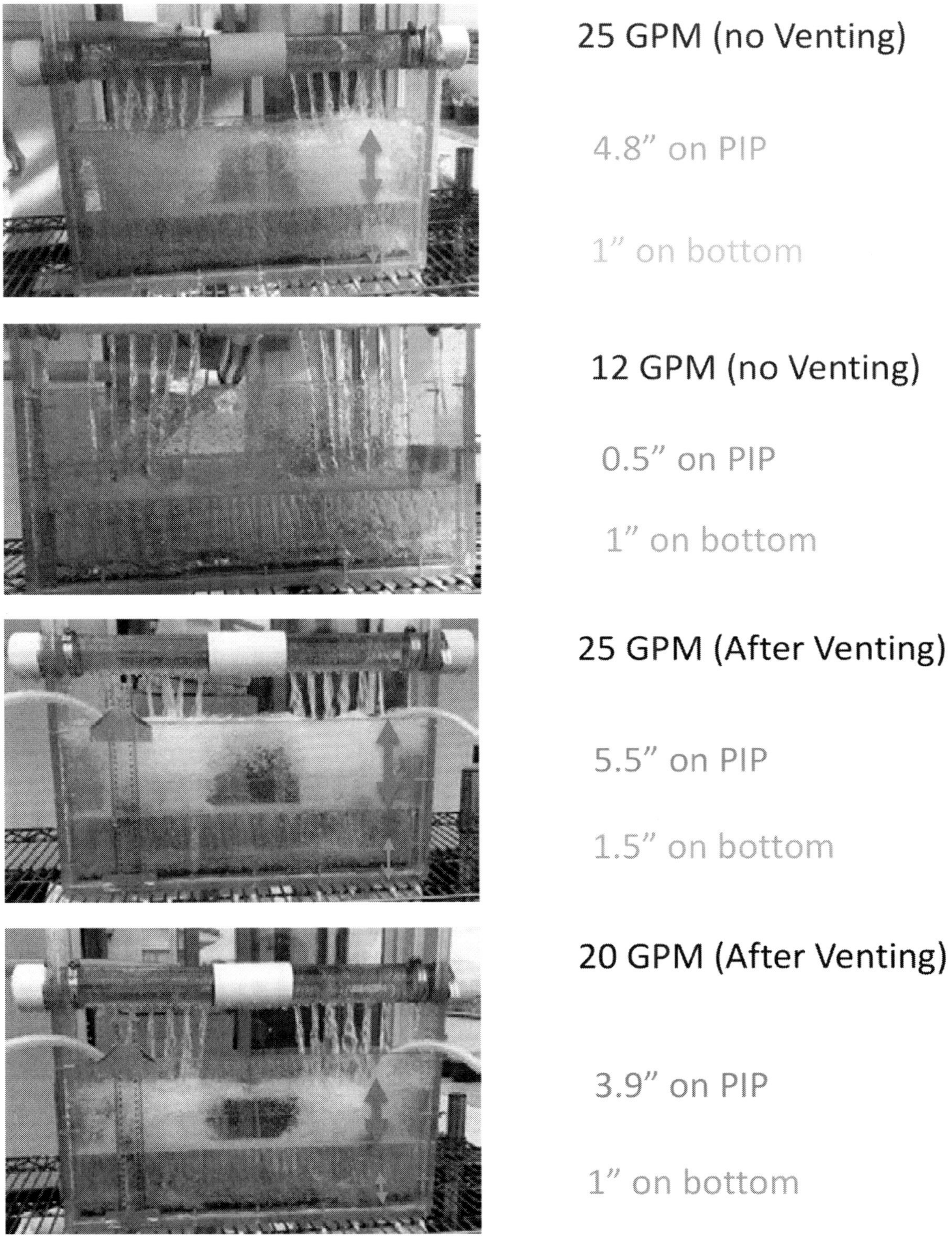

Figure 12. Tests with 10% open area PIP, 4” above the parting box floor, with and without venting the region under the PIP

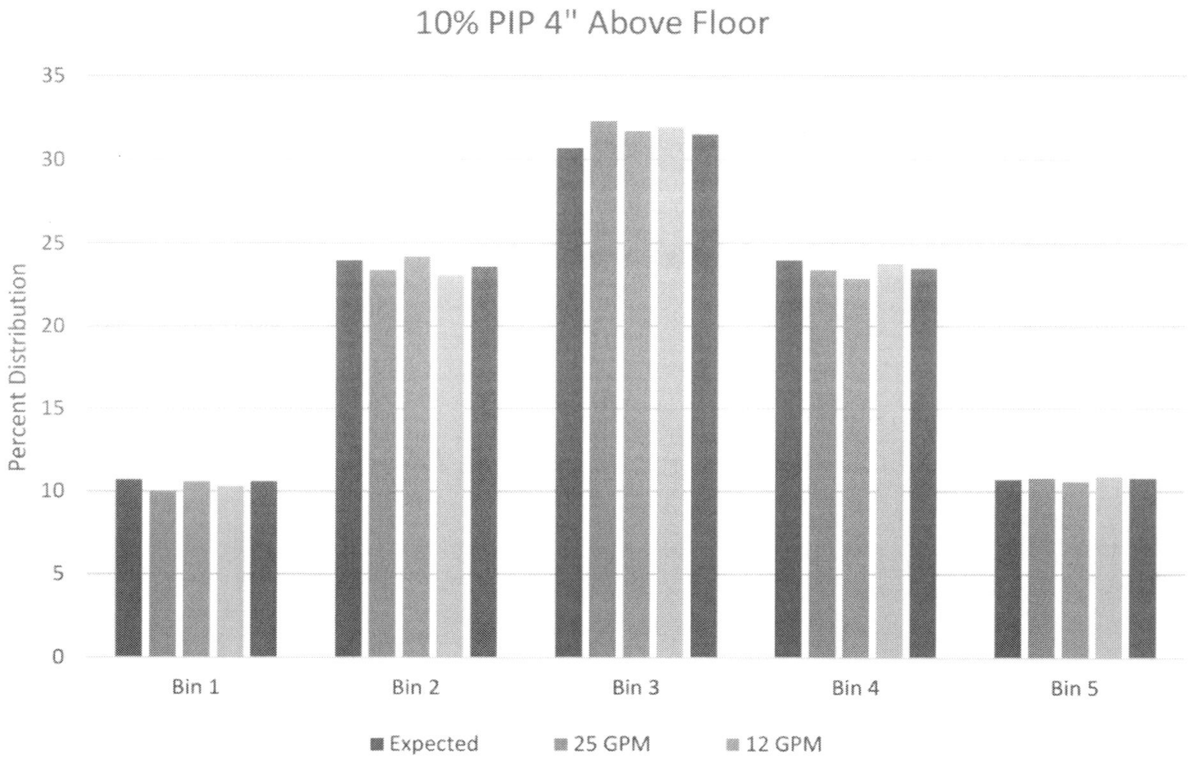

Figure 13. Liquid split among the bins compared to the desired split based of equal liquid flow rate through each perforation, with 10% open area PIP mounted 4" above the parting box floor.

25 GPM (After Venting)

6.5" on PIP

0.75" on bottom

20 GPM (After Venting)

4.1" above PIP

0.75" on bottom

12 GPM (no Venting)

0.75" on PIP

0.75" on bottom

Figure 14. Tests with 10% open area PIP, 2" above the parting box floor, with and without venting the region under the PIP

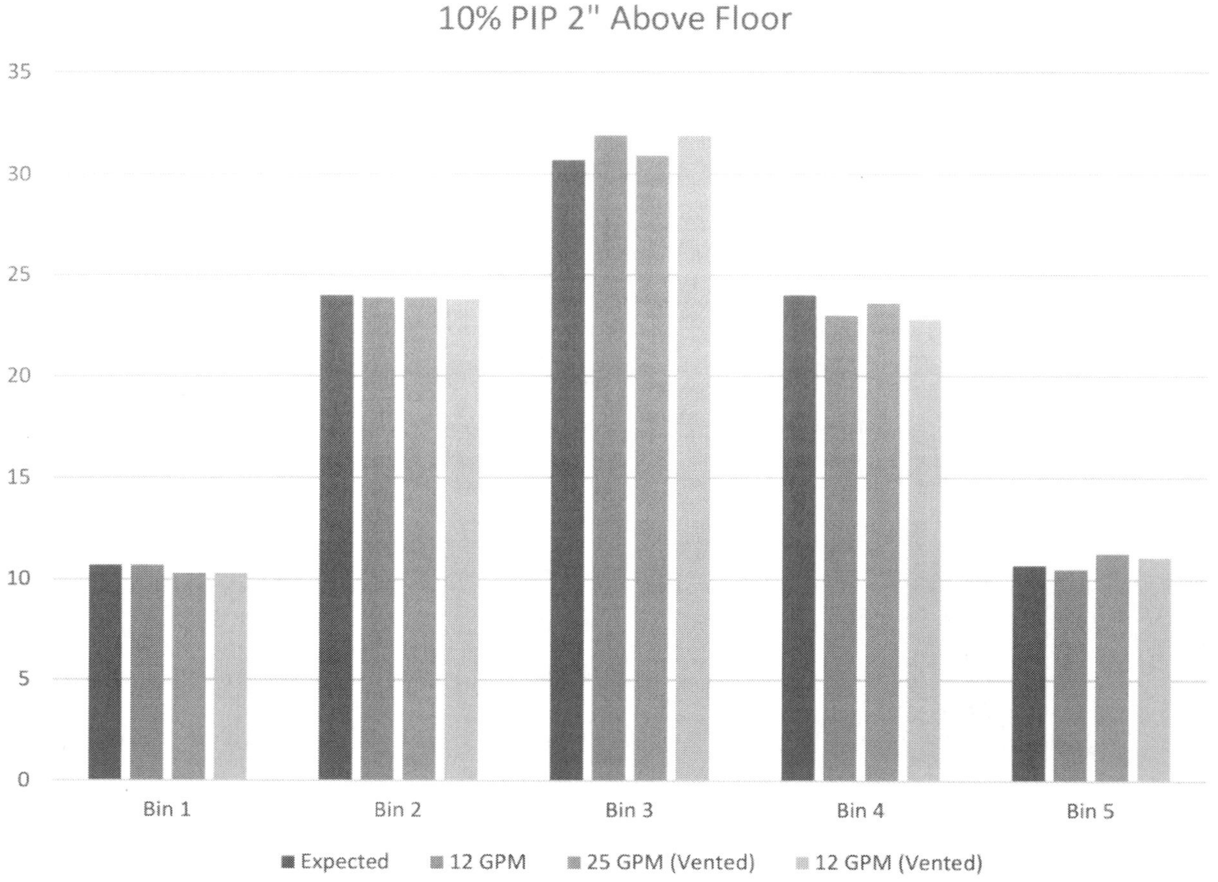

Figure 15. Liquid split among the bins compared to the desired split based of equal liquid flow rate through each perforation, with 10% open area PIP mounted 2" above the parting box floor.

Figure 16 compares the froth heights as determined visually from the photographs to the liquid heads calculated from the orifice equation (with coefficients per Perry's Handbook[6]) for the tests with the 10% open PIP Two solid curves are shown on Figure 11. The orange curve is the same as in Figure 5, based on the floor orifices alone. The purple curve also includes the liquid head calculated across the PIP. Since the PIP and the floor open areas are 10% and 5.8%, respectively, of the parting box cross section area, the additional liquid head incurred by the PIP is significant. The parting box froth height plotted in Figure 16 is the height above the parting box floor, not allowing for the PIP.

With the 10% PIP, it the froth heights were higher than the parting box liquid head calculated from the orifice equation (accounting for the PIP). The height difference was about 2 inches when the PIP was 4" above the tray floor, and about 1" when the PIP was 2" above the tray floor.

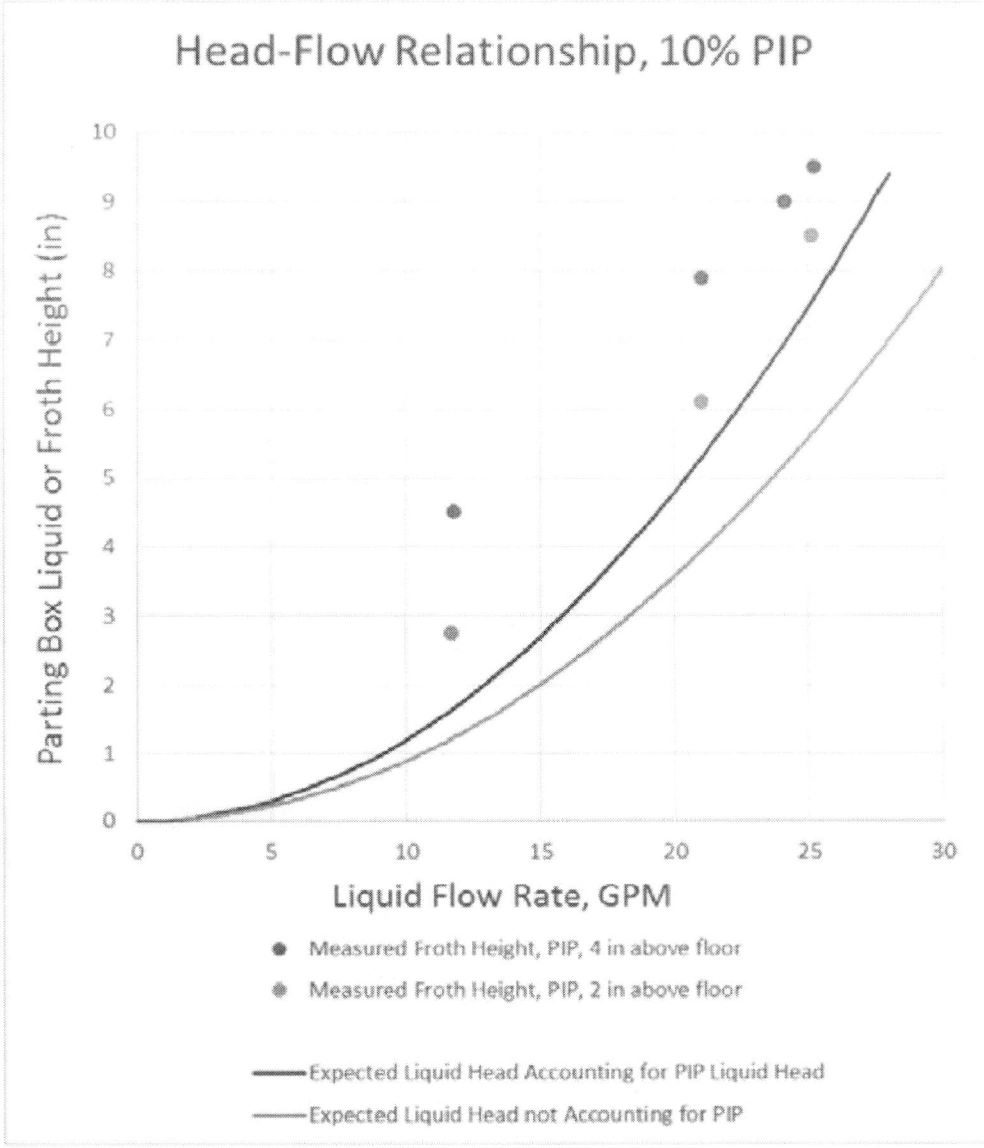

Figure 16. . Measured froth heights in the parting box compared to the head- flow relationship, 10% PIP tests.

Cause of the Air Gap under the PIP's

The unexpected air gaps under the PIP's prompted us to perform further experiments that identify the cause. One possible cause was that when the water first entered the parting box, air was trapped beneath the PIP and could not vent out. Another possible cause was that the air was being dragged in by the water jets and trapped by the water buildup on the PIP.

To check on these possibilities, a duplicate PIP with 10% open area was fabricated, this time equipped with two flexible 3/8" vent tubes as shown in the bottom two photographs on Figure 12 and the top two photographs on Figure 14. These vent tubes would vent the initial air. After a minute or two of operation all the initial air was expected to have been vented out, so the air gap beneath the PIP should disappear. Conversely, if the air was continuously dragged in by the water jets, the gap would have persisted and pressure would build up due to the water head on the PIP. This pressure could be measured simply by turning up the ends of the venting tubes and filling them by colored water that served as a manometer. So after the initial venting for a minute or two, the open ends of the venting tubes were turned upwards, forming a u-tube manometer on each side. The u-tubes were filled with colored water, and the difference between the two arms of each manometer was measured with rulers, giving a measurement of the pressure beneath the PIP.

In all tests, the vent tubes lowered the froth heights above the PIP by an inch or two, and raised the froth level below the PIP by about the same, but did not eliminate the gap. Following the venting, the manometer measured 4-5 inches of water at 25 gpm, declining to 2-3 inches at 20 gpm. This demonstrates that the gaps were induced by the water jets dragging air to the region beneath the PIP.

One concern we had is that the 3/8" tubes may not have been large enough to expel all the initial air. We therefore conducted one more experiment. We again used the PIP with the vent pipes, mounted 4" above the parting box floor, and ran the rig at 25 gpm, where the largest air gap was observed. Then we used the tubes as levers to tilt the PIP upwards, forming free venting space occupying about half the cross section area of the rectangular box. This would definitely have vented all the air initially trapped. Indeed, the gap completely disappeared. Then we allowed the PIP to reseat. Initially there was no gap, but within a few seconds the gap re-formed, and the froth heights returned to those measured previously. This experiment proved conclusively that the gap is formed due to trapping of the air continuously entrained with the water jets.

We performed one more test to verify this mechanism. With the 10% PIP plate mounted 4 inches above the parting box floor, we placed several bars of dry ice, large enough that the solids cannot penetrate through the 3/16" perforations, on the PIP. The rig was covered with a plastic sheet to prevent the wind from affecting the results. As the dry ice sublimed, we could watch its smoke being pushed by the liquid jets into the region underneath the PIP (Figure 17). This verified the postulated mechanism. We did not observe any smoke escaping down through the perforated parting box floor plate.

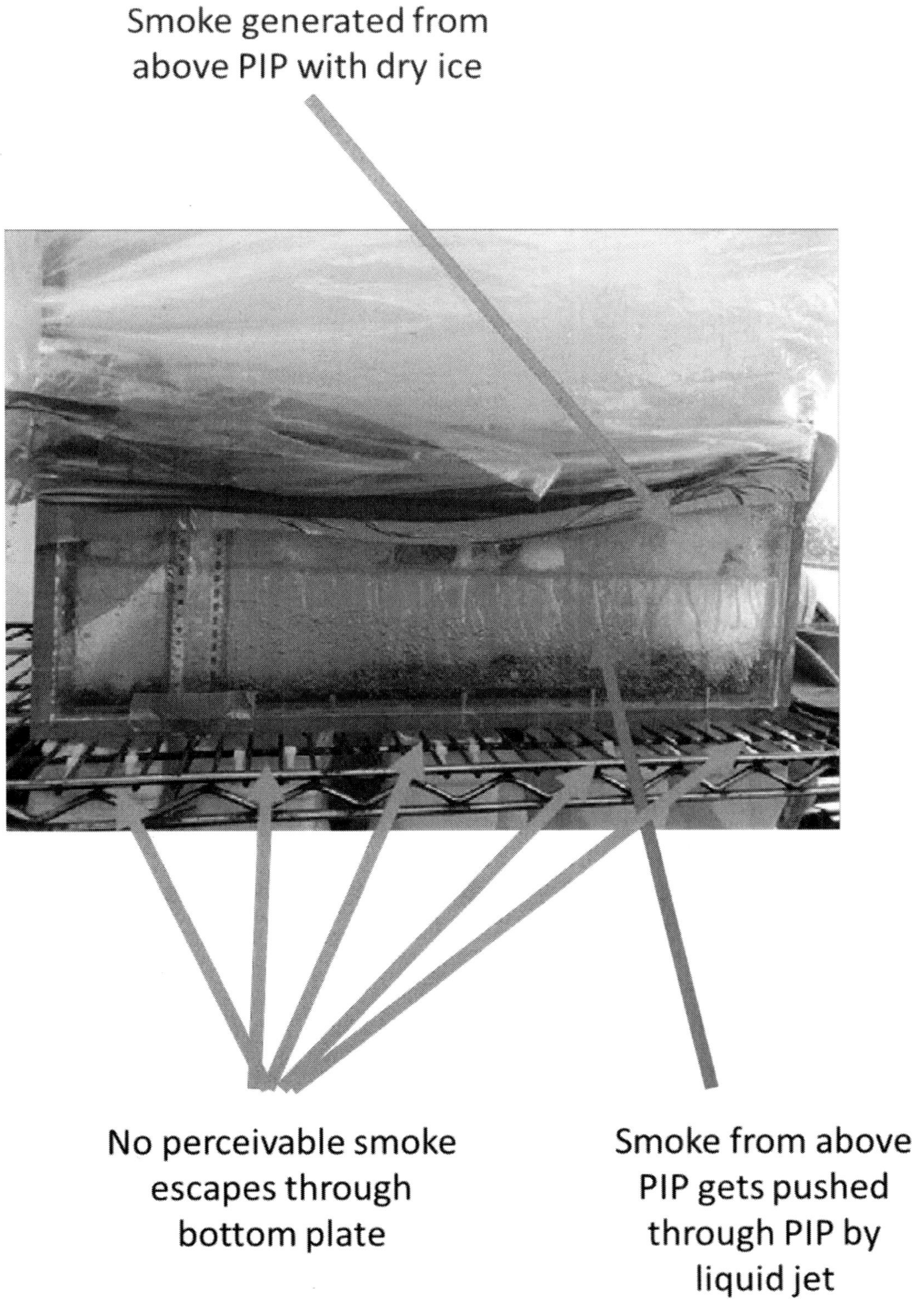

Figure 17. Smoke test, showing the cloud of sublimed dry ice being dragged by the liquid jets under the PIP but not under the perforated floor plate.

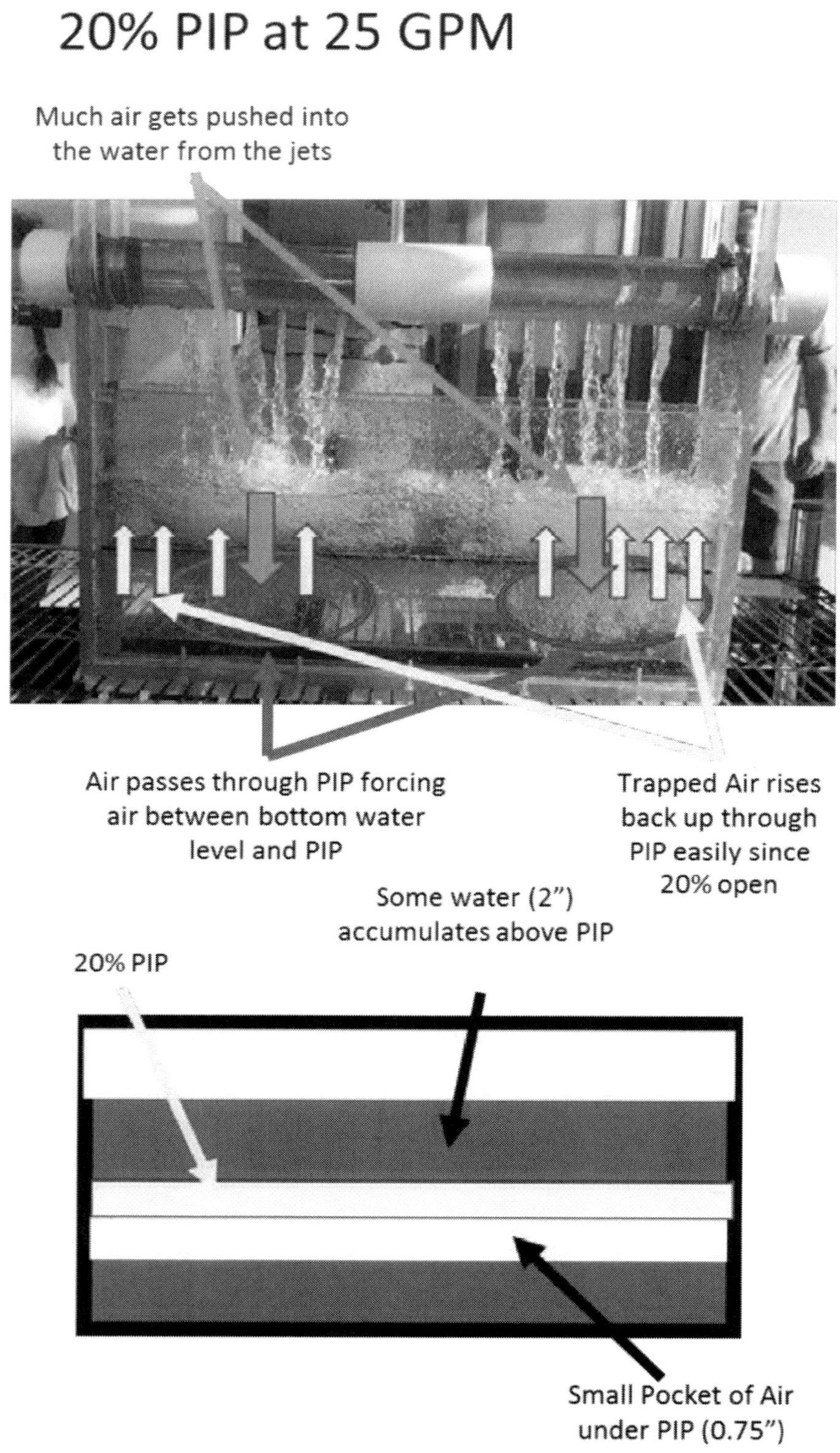

Figure 18. Mechanism of vapor gap formation beneath the PIP mechanism, 20% open area PIP, 4" above the parting box floor

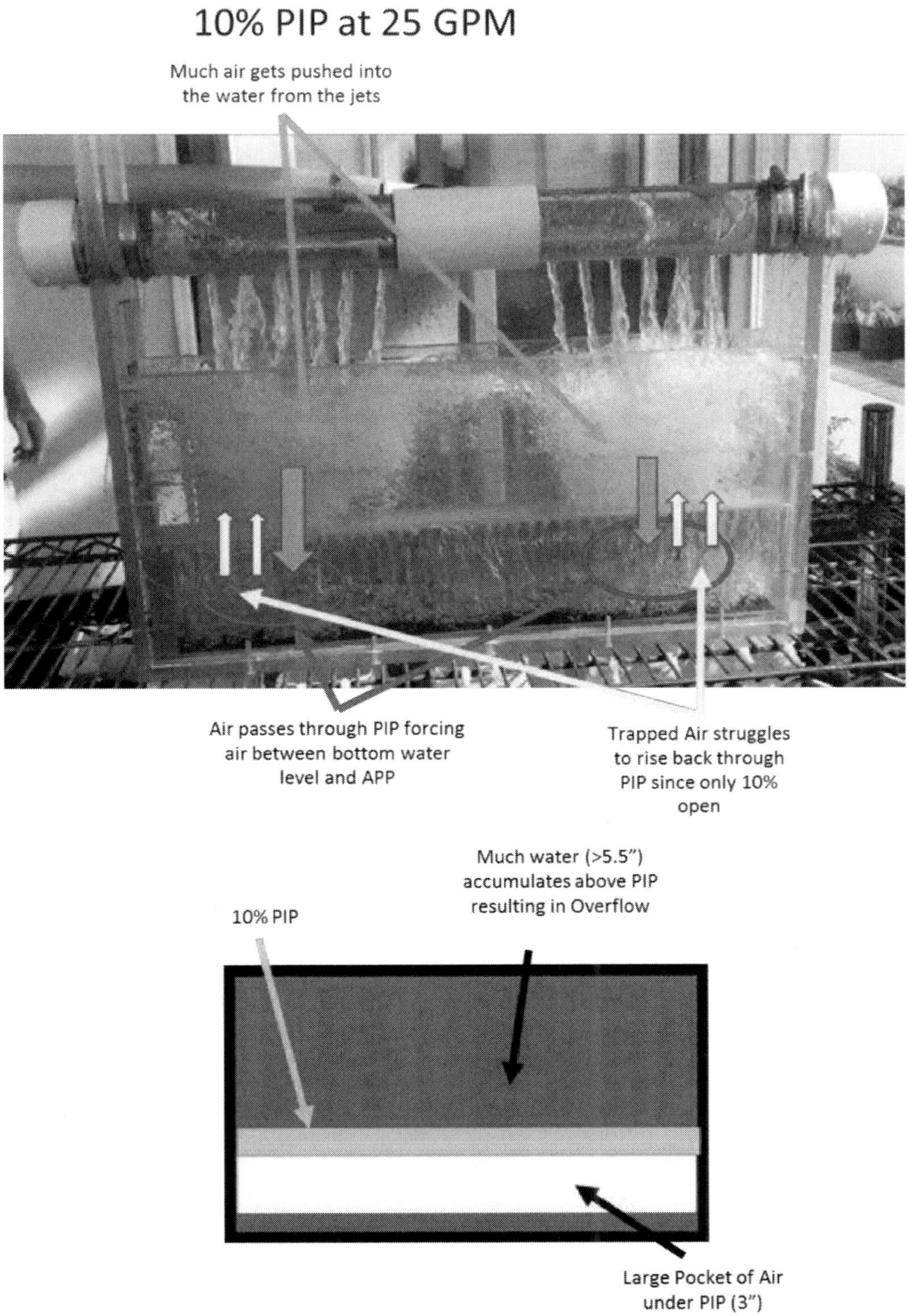

Figure 19. Mechanism of vapor gap formation beneath the PIP mechanism, 10% open area PIP, 4" above the parting box floor

Figures 18 and 19 illustrate the mechanism of air gap formation under the PIP. The water jets issuing from the feed pipe entrain air with them. This air passes through and accumulates beneath the PIP, being trapped by the prevailing water head on top of the PIP. The trapped air rises back relatively easily when the PIP has a large (20%) open area, as there is less froth backup (i.e., water head) on the PIP (Figure 18). The froth backup above the PIP and, consequently, the pressure buildup beneath the PIP are intensified with the more restrictive 10% open area PIP (Figure 19).

The liquid above the PIP imposes its head on the trapped air and helps maintain good liquid distribution at the outlet of the parting box. In all tests with the PIP's, an excellent to very good split of the liquid to the holes was measured, even though the liquid head above the floor plate was at times very low, sometimes as little as 1-2 inches or less. What was troublesome is that the presence of the PIP's generated a high froth buildup in the parting box and at high rates premature overflows. In an operating tower, with vapor ascending around the parting box, such overflow may result in entrainment and maldistribution.

Application to Other Systems

Water tests are a common technique for troubleshooting and evaluating distributors[3,7]. They have been highly effective for this purpose. We therefore expect our water tests findings to reliably extend to commercial systems. So all our findings, the frothing, liquid split quality, and air gaps are expected to broadly extend to systems using parting boxes of similar geometries and process conditions.

This said, there are many system-dependent parameters that may vary and are likely to impact the hydraulics of the parting box. For instance, the frothing can be much more intense in low surface tension or foaming systems, causing overflows even earlier. Parting box tests with non-aqueous systems will shed more light on these hydraulic issues.

Parting box and feed pipe geometries vary. There is a multitude of designs, even when a sparger pipe is used to feed the parting box. Different geometries produce different hydraulics. Testing different designs will broaden our understanding of the hydraulics and lead to developing reliable design guidelines for parting boxes.

Our tests placed some of the existing guidelines and practices under challenge. In particular, it showed that velocities as low as 6-7 ft/s of liquid exiting the feed pipe can cause intense frothing in the parting box that may be undesirable. The air gap phenomenon with PIP's, to our best knowledge first reported and explained here, adds a note of caution for PIP design. Premature overflows of froth above the top of the parting box have been observed due to frothing or due to air (or vapor) gaps below PIP's.

Conclusions

We performed water tests on a scaled down model of a parting box that exists in a hydrocarbon-separation tower that is unable to achieve the design separation efficiency. The parting box, both in the actual tower and in the tests, had a design that broadly complied with published good practice guidelines.

Our tests produced many unexpected results. They showed that in our tests, the liquid in the parting box was highly frothed up at both high and low liquid rates. The frothing in the parting box was due to water jets from the feed pipe entraining air into the water upon impact. This generated excessive froth heights in the parting box, and premature overflow of froth out of the parting box at high rates. The liquid split to the five clusters of holes at the floor of the parting box was good at the maximum rates but quickly deteriorated as the liquid rate was turned down.

Using a Perforated Impingement Plate (PIP) with either 20% or 10% open areas, mounted either 2" or 4" above the parting box floor, greatly improved the split of liquid to the clusters of holes at the floor of the parting box. This split was very good to excellent both at the maximum rates and at turndown down to about 50% of the maximum rates. Lower rates were not tested.

When the PIP's were used, an air gap formed beneath the PIP, with tall froth above, producing premature overflows at high rates, and low and relatively less aerated froth beneath the PIP. The air gaps resulted from the water jets issuing from the feed pipe entraining air with them. This air passed through the PIP and accumulated beneath the PIP, pressuring up the space between the parting box floor and the PIP. The trapped air rises back relatively easily when the PIP has a large (20%) open area, as the froth backup above the PIP is relatively small. The pressure buildup beneath the PIP and the froth backup above the PIP are intensified with the more restrictive 10% open area PIP.

The pressure buildup in the chamber beneath the parting box helps keep the liquid split out of the parting box good. What was more troublesome is that the presence of the PIP's generated a high froth buildup in the parting box and at high rates premature overflows at well below the expected overflow rates. In an operating tower, with vapor ascending around the parting box, such overflow is likely to generate entrainment and maldistribution.

References

1. Kister, H. Z. "What Caused Tower Malfunctions in the Last 50 Years", Trans. IChemE Vol. 81, Part A, January 2003. Glitsch, Inc., Bulletin 4900, 6th Ed., Dallas, Texas, 1993.
2. Lockett, M. J., and J. F. Billingham, "A Simple Method to Assess the Sensitivity of Packed Distillation Columns to Maldistribution", *Trans. IChemE* **80:** Part A, p. 373, May 2002.
3. Olujic, Z., H. Jansen, "Large-Diameter Experimental Evidence on Liquid Maldistribution Properties of Structured Packings", *Chem. Eng. Res. Des.* **99:** p. 2, 2015.

4. Kunesh, J. G., L. L. Lahm, and T. Yanagi, "Commercial Scale Experiments that Provide Insights on Packed Tower Distributors", Ind. Eng. Chem. Res., 26, 1845, 1987.
5. Moore, F., and F. Rukovena,"Liquid and Gas Distribution in Commercial Packed Towers", Chem. Plants and Processing (European edition), p.11, August, 1987.
6. Kister, H. Z., P. Mathias, D. E. Steinmeyer, W. R. Penney, V. Monical, and J. R. Fair, "Equipment for Distillation, Gas Absorption, Phase Dispersion, and Phase Separation", Section 14, in R. H. Perry and D. Green "Chemical Engineers' Handbook", 9h Ed., McGraw-Hill, N.Y., 2018.
7. Olsson, F.R. "Detect Distributor Defects Before They Cripple Columns", Chemical Engineering Progress, p.57, October 1999
8. Kister, H. Z. "Distillation Operation", McGraw-Hill, N.Y., 1990.
9. Maćkowiak, J. "Fluid Dynamics of Packed Columns", Springer-Verlag, Berlin Heidelberg, 2010.
10. Norton Company, "Packed Tower Internals", Bulletin TA-80R, 1974 (available from KochGlitsch, Wichita, Kansas).
11. Fadel, T. M. "Selecting Packed-Column Auxiliaries", Chem. Eng., January 23, p.71, 1984.

551715 Development of Structured Packings Using Additive Manufacturing and 3D Printing

Additive manufacturing (3D printing) has been assessed for the development of structured packings for humidity control at a spray dryer inlet. The desired air throughput for this application is a superficial velocity of 3.92 m/s. Two packing types were created using 3D printing, a layered, structured Pall ring, and a twisted sieve tray design. Experimental results were analysed to evaluate the throughput and mass transfer potential of the two packings. Whilst the structured Pall ring packing had a high mass transfer efficiency (NTU up to 1 for a packed height of m), the column flooded at gas velocities of 2.7 m/s and was hence unsuitable for the desired application. The pressure drop of the twisted sieve tray packing (420 Pa/m) was determined to be higher than that of Mellapak (700 Pa/m) at gas and liquid loadings of F_v=1 m/s(kg/m^3)$^{0.5}$ and q_L = 5 m^3/m^{2}h, respectively. However, the twisted sieve tray packing did not flood at the required gas flow conditions and so had a suitable capacity. The twisted sieve tray packing achieved NTU up to 0.6 for a packed bed height of 0.2 m at the desired superficial velocity. Compared with industrial structured packing designs, such as Mellapak (250X), the structured Pall ring packing had a lower HTU, indicating better mass transfer performance than both the Mellapak and twisted sieve tray packing. The use of 3D printing allowed the timely and reliable design and construction of the two structured packings.

551716 Improving Mass Transfer and Hydraulic Performance of Random Packings By a Non-Traditional Approach

The chemical process industry benefits greatly from any improvements on mass transfer equipment performance due to its impact on energy consumption in distillation intensive processes containing conventional equipment -trays and/or packings-. In existing grass-roots plants, processes intensification - higher throughputs, higher purities, or yields- is typically achieved by upgrading from conventional to high-performance equipment and operating conditions. In high pressure distillation systems designed with trays, these are upgraded to either high-capacity trays or high-performance random packings. The latter typically achieved using fourth-generation random packings. These types of packings exhibit better hydraulics that leads to higher efficiencies as a consequence of an improved geometry.

The non-conventional approach presented in this work consists in blending two-different sizes of random packings. This approach has been available for 15 years. The random packing blend capacity is like the larger random packing size with an improved efficiency. This efficiency is close to the smaller packing size, or 10-15% higher efficiency of the larger packing.

In this work, blending two different size of third-generation random packings will exhibit hydraulic and mass transfer characteristics like those of fourth-generation packings.

Experimental work conducted in a state-of -the-art test system with air-water and a standard distillation system confirm this non-traditional approach. The hydraulic modeling of these random packing blends confirms a well-known design methodology for these commercial random packings.

Case studies will be presented.

549795 Numerical Simulation of Two-Phase Flow and Interfacial Species Transfer in Structured Packings

Abstract

To optimise the separation performance of structured packings in distillation columns, a thorough understanding of the fluid dynamics as well as the transport processes within these column internals is vital. Conversely, this means that only simulations comprising both phenomena allow a comprehensive analysis of the process. This study presents the setup to simulate the fluid dynamics, the GCST model to consistently incorporate the interfacial species transfer and the combined setup which merges both fields into a single simulation run for determination of the separation efficiency.

Introduction

Distillation columns have a wide application in many sectors of process industry. By them it is possible to separate the individual components of a mixture. This is done by multistage continuous distillation where the separation is enabled through evaporation and condensation.

To optimise the efficiency of this process the interface between the gas and the liquid phase must be maximised, whereas the pressure drop should be minimised. Both requirements are met by so called structured packings. Against the background of further improvement, a thorough understanding of the fluid dynamics and transport processes inside the mentioned column internals is vital. To approach this target, numerical simulations of single- and multiphase flow inside structured packings, including species transfer, are conducted and closely evaluated.

Simulation of fluid dynamics in structured packings

In a first step the fluid dynamics in structured packings are investigated and the results achieved are presented in Hill et al. (2016). In the course of spatial modelling the smallest possible periodic domain could be identified which in combination with cyclic boundary conditions is an efficient calculation region. First, the case of single-phase gaseous flow is analysed, focusing on the visualisation and understanding of the transient flow structures occurring at higher flow rates. Second, a two-phase flow simulation is conducted to identify the local distribution of the liquid phase whereby the results are compared to experimental data.

GCST model to simulate interfacial species transfer

To enable the simulation of interfacial species transfer in the context of finite volumes, a new model is derived and validated. The model is based on the Continuous Species Transfer (CST) method introduced by Marschall et al. (2012) and Deising et al. (2016), which is generalised in a way that not only absorption but also distillation can be simulated and is therefore termed Generalised Continuous Species Transfer (GCST) model. The derivation and validation of the model is published in Hill et al. (2018).

Combined setup for simulation of distillation in structured packings

In this last step the two previously described fields of research were combined and two-phase flow and species transport in structured packings are simulated simultaneously. Before doing so new boundary conditions simulating the head and the bottom of a column had to be developed. The conducted simulations cover different liquid and gas loads and as a parameter to evaluate the results the Height Equivalent to a Theoretical Plate (HETP), a parameter describing the separation efficiency, was calculated.

Conclusions

Combining the GCST model with a 3D transient direct numerical simulation of the fluid dynamics in a structured packing enables the prediction of the separation performance (HETP) of the packing. Hereby, only physical properties are required, and no fitting parameters are part of the model. In-depths analysis of the simulation data will not only help to optimise packing geometries but also to gain a deeper insight into the process itself. In addition, important parameters like the specific effective surface area, degree of wetting or mass transfer coefficients can be extracted.

Acknowledgments

The authors gratefully acknowledge the Gauss Centre for Supercomputing e.V. (www.gauss-centre.eu) for funding this project by providing computing time on the GCS Supercomputer SuperMUC at Leibniz Supercomputing Centre (www.lrz.de).

References

1. Deising D., Marschall H., Bothe D., A unified single-field model framework for Volume-Of-Fluid simulations of interfacial species transfer applied to Bubbly Flows, Chemical Engineering Science, Vol. 139, pp. 173-195 (2016)
2. Hill S., Acher T., Hoffmann R., Ferstl J., Rarrek A., Klein H., CFD Simulation of the Hydrodynamics in Structured Packings, Conference Paper, 9th International Conference on Multiphase Flow, Florence (2016)
3. Hill S., Acher T., Hoffmann R., Ferstl J., Deising D., Marschall H., Rehfeldt S., Klein H., Numerical Simulation of Two-Phase Flow and Interfacial Species Transfer in Structured Packings, Chemical Engineering Transactions, Vol. 69 (2018)
4. Marschall H., Hinterberger K., Schüler C., Habla F., Hinrichsen O., Numerical simulation of species transfer across fluid interfaces in free-surface flows using OpenFOAM, Chemical Engineering Science, Vol. 78, pp. 111-127 (2012)

551720 Utilizing Gamma Scans to Monitor Fouling Accumulation in Packed Towers

In some processes, a common problem is the accumulation of fouling material in the packed beds of separation towers. Fouling can cause conditions such as liquid and/or vapor maldistribution and excessive liquid holdup in packed columns, resulting in a loss of efficiency, an increase in operating pressure drop and premature flooding. Fouling may result from foreign material entering a column or from polymerization or decomposition occurring inside the column. The fouling may start under normal conditions or begin to build due to abnormal operating conditions.

During a turnaround, a visual inspection can be performed on the upper and lower layers of packing, but unless the packing is removed, it is impossible to see fouling material within the bed. In some cases, towers with fouled beds are restarted without cleaning or replacing the packing, because no fouling was visually seen on the upper or lower layers of packing during the inspection. This can ultimately lead to costly unplanned outages.

Gamma scanning is a non-intrusive and cost-effective way to monitor a packed tower for the accumulation of fouling material. Many operators implement a routine maintenance program where the column is periodically scanned to monitor the progress of the condition over time. This type of monitoring program typically begins with a baseline scan of the column.

This paper will provide case studies that illustrate the effectiveness of utilizing gamma scans to monitor fouling accumulation in packed towers.

Common Refinery Distillation Design Errors

by

Elizabeth Lieberman and Norm Lieberman
Process Improvement Engineering
Metairie, Louisiana
norm@lieberman-eng.com

Abstract

Based on the authors' refinery process troubleshooting experience, common process design errors have been extracted from their troubleshooting seminar outline. These errors range from the B.P. Texas City Raffinate Splitter Explosion, to a steam reboiler bottleneck that the authors corrected recently at the HPCL refinery in Mumbai.

I – Sour Water Stripper

Mumbai, India – Reboiler Balance Line

Inadequate steam stripping was causing NH_3 in stripper bottoms to exceed 20 ppm needed for crude desalter wash water. The problem was lack of steam flow to the reboiler. Field investigation revealed the problem. The balance line connecting the channel head to the condensate drum was connected to the wrong side of the channel head pass partition baffle. While it is industry design practice to connect the balance line to either steam supply line, or to the channel head above the horizontal pass partition baffle, this practice is <u>wrong</u>.

In this case, we measured a DP of 0.1 BAR (three ft. of water) on the steam side of the reboiler. Which meant that the condensate level in the channel head had to be at least 3 ft. higher than the condensate level in the condensate collection pot.

However, the bottom of the condensate pot was only about 2 ft. below the bottom of the channel head. The top of the pot was higher than the top of the channel head. If steam flow to the reboiler was increased by 41%, then the level in the channel head would have increased from:

- $3' \times (1.41)^2 = 6'$

This would have submerged most of the tubes in the reboiler in water, and thus limited the stripper's reboiler duty.

LESSON – Connect the condensate pot balance line to a ¾" valve immediately beneath the lower pass partition baffle in the channel head (Fig. I).

II – Elevating Overhead Condenser

Perhaps the most common problem we have seen in distillation towers, is excessive elevation of the overhead, water cooled, condenser. The common assumption, is that the elevation of a water cooled condenser is not critical. The basis for this misconception is that if the cooling water rises, and loses head pressure, head pressure will be regained as water flows back down the cooling water return line.

It is true that the lost head pressure is restored as the cooling water flows down to the water return header. However, at the top of the loop, at the condenser water outlet nozzle, the water will be at a sub-atmospheric pressure. Dissolved air will then flash out of the heated water, and restrict water flow (Fig. II).

LESSON – Do not allow the heated cooling water return from an elevated condenser outlet, to drop one or two psi below atmospheric pressure.

III - Distillation Tray Assembly

To promote tray failure, we recommend the "Washer" assembly shown below. In this type of assembly, two tray sections are squeezed together between two washers. Only one of the two tray sections has a hole. A bolt is passed through this hole, and through the larger washer. The adjacent tray panel is slipped under this large washer. Then, the assembly is secured with a nut screwed onto the end of the bolt.

Such an arrangement is easy to install, as no holes have to be aligned in the adjacent tray panels. Unfortunately, a pressure surge can easily pull the tray panels apart. Which may be good, if you're selling replacement trays, but bad if you're trying to run a Delayed Coker Fractionator in Texas City. This problem is enhanced if you have purchased 14 gauge, rather than 10 gauge, trays. The better method of tray assembly is the "Back-to-Back" tray – i.e., both.

Edges of the tray panel are bent down to form half an I-beam. Then, the two vertical edges are thru-bolted together, with the bolts inserted in a horizontal plane, to form an integral I-beam, connecting the two tray panels (Fig. III).

A short piece of ¼" steel bar, is then bolted to the edge of each I-beam, and slipped under the tray ring – but not welded to the ring.

Tray panels are constructed from 10 gauge – not thinner 14 gauge material.

LESSON – Tray mechanical details must be specified by the refiner; not left to the fabricator.

IV – Sour Water Stripper Design

The typical SWS design:

1. Preheats feed water from 90°F to 200°F.
2. Five top reflux trays.

Both serve no purpose. In 1969, I designed a sour water stripper at Milford Haven (U.K.) with neither feature and it worked fine. Two years back, we ran a field test by-passing the feed preheater in India with no increase in the NH_3 of SWS bottoms.

As the stripper is run on total reflux, reflux serves no purpose, except to return excess condensate to the tower. The water content of SWS NH_3 gas to the SRU is controlled by the reflux drum temperature, not the tower top temperature (Fig. IV-A).

Preheating the feed does not save steam. The reboiler steam is needed to effect a stripping factor in the 10 – 15 trays above the reboiler. Feed preheat only generates incremental reflux for the useless reflux trays (Fig. IV-B).

LESSON – Just because SWS feed preheat and reflux trays are standard in the industry, does not make it right.

V – Vertical Baffle in Tower Bottoms

For towers with a vertical baffle in the bottom of the tower, the reboiler return should be placed on the hot side of the tower (on the side from which bottom product is drawn). Not on the cold side (on the side that the bottom tray overflows into) (Fig. V).

Much of a tower's fractionation takes place in the reboiler. Returning reboiler effluent to the cold side of the vertical baffle, causes the liquid from the bottom tray to spill directly into the hot side of the vertical baffle. This by-passes the bottom theoretical stage of the tower.

We substantially reduced the ethane content of an FCU gasoline stripper bottoms in Port Arthur by making this change.

LESSON – The industry-wide practice of routing the reboiler return flow to the cold side of the vertical baffle, and overflowing the baffle onto the hot side, is inherently wrong.

VI – Chimney Tray Overflow Pipe

In Mumbai, India, there is a large FCU designed by an American contractor. The debutanizer, which initially worked well, flooded after a few years. Chemically cleaning the tower trays did not help. I corresponded with my friend SriGanesh for six months without solving the problem. However, after being on-site for 20 minutes, we recognized this familiar error. That was, lack of provision on a total-trap-out chimney tray for internal overflow.

Actually, the chimney tray did have provision for internal overflow. As the reboiler fouled on the shell side (which was not cleaned when the trays were cleaned), the liquid level on the chimney tray backed-up. But, the top of the overflow pipe was 18" above the bottom tray seal pan (Fig. VI).

LESSON – When detailing tower internals, lay out the vessel sketch on graph paper, to scale, with a pencil. One inch per foot. Ask yourself: Does this look right?

VII – Raffinate Splitter Explosion

BP Texas City

This tower dated from the mid-1970's, when I worked for AMOCO Oil. I assisted in its design. Gary Elmer, ten years my senior, was the lead process engineer. We did not specify whether to route the relief valve to an atmospheric vent, or to the plant flare. As a result, the relief valve was routed to an atmospheric vent. Thus, when the relief valve opened, a cloud of naphtha enveloped a trailer, detonated, and 15 people were killed.

BP's investigation concluded that was the fundamental cause of the explosion. The government investigation and litigation about the incident, cost BP $4,000,000,000.

Certainly, venting the relief valve to an atmospheric blow-down stack was wrong. Of equal import, was draining the blow-down stack to a pressurized collection header (i.e., the 3# sewer), via a loop-seal, that prevented the blow-down drum draining due to "Vapor Lock" at the top of the loop (Fig. VII).

But, what caused the splitter relief valve to open in the first place?

Certainly, because of confusion in the operating staff, as to how their level indication worked, the tower was filled with raffinate, which overflowed into the vapor line. However, this did not pressure the tower above its relief valve set-point. The problem was that the relief valve was not located at the top of the tower. But about 70' below the tower top.

When the naphtha overflowed the tower, it exerted a head-pressure on the pressure relief valve. Which then caused the valve to open. The valve had been located in the pipe rack (about 20' above grade) for ease of maintenance, and to reduce piping costs.

LESSON – Place PSV's at the top of the tower itself. On the vessel sketch, indicate PSV nozzle location and size. Don't leave it to the mechanical engineers.

VIII – Crude Tower Top P/A

One of the world's largest refineries is located in eastern Canada. To enhance Top Pumparound capacity, the designer of their crude unit, decided to use structured packing for TP/A with a crimp size of ½". They also made the packing out of 410 steel. These dual mistakes off-set each other. The packing corroded away before it could salt-up, plug, and flood (Fig. VIII).

Eventually, the refiner decided to replace the packing with an alloy that resisted HCl corrosion. The packing then progressively flooded after a few months as it fouled, but remained intact.

LESSON – Don't employ engineering contractors that don't have operating experience. The upper trays of crude towers, should be grid trays, with a "lift" of ½".

IX – Excessive Thermosyphon Circulation

A Benzene-Toluene fractionator near our home in New Orleans was subject to flooding. The problem was excessive thermosyphon circulation. The design error was the location of the reboiler return nozzle very close to the bottom tray, and above the bottom tray seal pan.

When the operators increased reflux, they found that they also had to increase hot oil flow to the reboiler, to keep the reflux drum from going empty. The increased reboiler duty, increased rates of thermosyphon circulation, which splashed violently against the seal pan and bottom tray deck. Which caused the tower to flood (Fig. IX).

We tried to lower the tower level, but this caused the bottom product pump to cavitate due to loss of NPSH. As there were no valves in the reboiler circuit, the flooding problem and consequent loss of fractionation efficiency could not be corrected without a shut-down to install an impingement plate in front of the reboiler return nozzle.

The reboiler return nozzle should be located a few feet below the bottom tray seal pan, not a few inches below the bottom tray deck.

LESSON – Design features intended to reduce column height by a few feet, can result in millions of dollars in degraded product values.

X – Tray Hydraulics

1 - Sour Water Stripper – Indonesia

We were troubleshooting a poorly performing sour water stripper for a large refinery in Indonesia. That is, excessive NH_3 in stripper bottoms.

The tray data sheet stated the following values:

- Weir Ht = 3"
- Crest Ht = 2"
- Delta P dry = 0.3"
- Tower I.D. = 8'-6"

Trays were conventional valve trays, that are in widespread use in the refining industry – except never used in towers we retrofit, or design.

We explained to our Indonesian colleagues that the trays were leaking, and destroying tray fractionation efficiency. We explained:

- Wt. of liquid on trays = crest ht + ½ (weir ht) = 3½", which was 10 times larger than the vapor DP of 0.3" (Fig. X).

"But, Lieberman," they objected, "these are valve trays. The valves will close to retard tray leakage."

We responded, "Valve trays are a fraud. Only vapor DP supports the weight of liquid on the tray."

"But, Lieberman," they objected, "even if the trays leak, as long as the trays are level, vapor-liquid contacting should not be impeded."

"Level? This is not a 36" I.D. pilot plant being operated in California by my pal Henry Kister. This is a real world, large I.D. tower – where trays are never level. DP dry must be 50% minimum of the crest height plus half of the weir height (to account for foam) to ensure good vapor-liquid contacting and tray fractionation efficiency."

LESSON – If you want to be able to operate at variable low vapor rates, use Bubble Caps.

2 – Crude Tower Bottom Stripping Tray Retrofit

A Canadian refinery retrofitted the bottom six steam stripping trays on their crude tower. The objective was to increase the number of trays. As part of the retraying project, design downcomer clearances were reduced from two inches to one inch. While the calculated head loss under the downcomer was not excessive, the trays flooded at less than design flows.

LESSON – It's our experience that using downcomer clearances of less than 2", is looking for trouble.

XI – Crude Tower Bottom Stripping Tray Retrofit

A refinery in Eastern Europe retrofitted the two-pass steam stripping trays in the bottom of the crude tower but did not replace the two existing seal pans. The problem was seen by the operators as premature flooding and causing them to reduce the feed rate, to the crude tower.

The refinery had the tower scanned so as to help discover the source of the problem with the stripping trays, and found that the flooding was starting on the lowest tray. When we inspected this tower at the turn-around we found that the horizontal clearance between the down comer and the lip of the seal pan had been reduced from 2 inches to 1 inch.

When considering down comer clearance for a Seal Pan, it is just as important to have sufficient horizontal down comer clearance as vertical down comer clearance (Fig. XI).

LESSON – When planning to retrofit parts to existing equipment, take care to provide precise measurements of existing equipment to the retrofit equipment suppliers. Take care with installation to ensure that there are adequate liquid and vapor pathways.

XII – Vacuum Tower Flash Zone Pressure

In the center of Spain, there is a refinery dating from World War II, designed originally to convert coal to oil. The refinery had a vacuum tower that was running with a flash zone pressure of 40 mm Hg and a tower top pressure of 30 mm Hg. By making operational changes to the ejectors, we reduced tower top pressure to 14 mm Hg. But the flash zone pressure only dropped to 34 mm Hg. Quite a disappointment!

A detailed DP survey showed that half of the observed new 20 mm Hg tower pressure drop, was across the LVGO chimney tray. We calculated the DP using an orifice coefficient of 0.9 and confirmed the observed 20 mm Hg DP.

The problem was that the chimney open area was only 22% of the tower area. The open area between the chimney hats must be approximately the same as the open area of the chimneys, and additional area is required for:

- Center draw-off sump.
- Hat overlaps of chimneys.
- Tray support ring.
- Edge losses.
- Chimney wall thickness.

As a result, the maximum chimney tray open area can be only about 36% of the tower area.

We offered to design a new LVGO chimney tray that would reduce the DP from (Fig. XII):

- $(10) \bullet (22\%/36\%)^2 = \underline{3 \text{ mm Hg}}$

But to date, we're still waiting for the P.O. number to appear in our email.

LESSON – Tower internal details should be specified by the process design engineer and not left to a 3rd party.

XIII – Level Tap Location

A. Union Carbide produced NutraSweet at their plant in Augusta, Georgia, in the early 1980's. A methanol stripper was flooding and reducing stripping efficiency. The problem was that the top level tap was located 2 ft. above the reboiler return nozzle. As this was a forced (i.e., pumped) circulation reboiler, any bottoms liquid level held above the reboiler return nozzle, would force the bottoms level up against the bottom tray, which would flood, and progressively cause the entire tower to flood. I reduced the level set-pt. from 60% to 30%, to correct this design error (Fig. XIII-A).

B. We had designed a vacuum flasher for CO_2 removal from wellhead waste water in Colombia. We had located the level taps six feet below the packed bed and at the bottom tangent line.

However, when we went to the site to start-up the flasher, we found that the top level tap was actually located halfway up the packed bed, and the bottom tap 6" below the bed.

So, we abandoned level control, and switched to bottom pump suction pressure control, using the existing pump suction pressure transmitter, to control the former LRC control valve at the pump discharge (Fig. XIII-B).

C. A heavy coker gas oil stripper had an erratically high liquid level indication. We could see droplets of water falling through the yellow oil in the gauge glass and hear water gurgling, as it hit hot oil at the bottom level tap, and saw that it flashed back to steam. We could also see steam blowing back in the gauge glass, creating an erratically high liquid level in the glass. On the control panel, the level indication was also erratically high.

The problem was that the top level tap was located just a few inches above the stripping steam inlet nozzle, and on the same side of the tower. It should have been located on the opposite side of the tower, two feet below the steam inlet nozzle (Fig. XIII-C).

To correct this problem, we connected a nitrogen purge through a 1/16" restriction orifice to the top of the gauge glass. The small N_2 flow prevented the flow of steam into the gauge glass assembly, and the level-trol column. The indicated level stabilized on the control panel, and in the local level glass.

<u>LESSON</u> – A review of the final vessel sketch, including locations of level taps, and their relation to process nozzles, would prevent these problems. In the example from Colombia, we failed to insist on a review of the vessel sketch, after it had been completed by mechanical designers.

XIV – Crude Tower Overhead

Towers with a two-stage overhead system (hot drum and cold drum) are commonly used in crude distillation service. The purpose is to prevent condensation of water in the hot drum.

The hot drum provides a large portion of the crude tower's fractionation ability. It serves as the top theoretical stage of the crude still. Field tests show it provides as much fractionation as the top 3 trays in the tower (Fig. XIV).

Using the reflux line from the cold drum greatly diminishes fractionation efficiency of the two-stage overhead system. We experimented with this concept at the former Coastal Refinery in Aruba, where we monitored naphtha vs. jet fractionation efficiency.

<u>LESSON</u> – Never use cold reflux on a crude unit two-stage overhead system. This practice also promotes corrosion and fouling both in the tower, and the first stage condenser.

XV – Using High Pressure Steam in an FCU Gasoline Splitter Reboiler

Down river from our home, along the Mississippi, is a refinery with a hydrotreated FCU gasoline splitter. This fractionator was subject to flooding at only 60% of its design capacity.

"Mr. Norm, every time we get a high level alarm in the reflux drum water boot, the splitter floods," observed the panel operator.

"Henry, that's 'cause you're refluxing water down the tower. Where's all that water coming from?"

"Wet feed, I reckon, Mr. Norm."

So I went out, got a feed sample and let it cool off to 60°F. The sample had not a trace of water or haze.

As the overhead condenser was an air cooler, the only possible source of water was a steam leak in the reboiler.

This tower had a 400 psig steam reboiler. The tower pressure was about 40 psig. Actually, 100 psig would have been sufficient to supply heat to the reboiler, especially since the reboiler was also massively oversized.

From the level I could see in the condensate pot, it seemed that 90% of the tubes were full of water at a pressure of 400 psig. There was no control valve on the steam inlet. The flow of steam was controlled by the condensate outlet valve (Fig. XV).

Even a tiny tube leak, submerged in water, with a 360 psi (400 psig – 40 psig) differential pressure, would cause a large flow of water into the reboiler vapor return line.

LESSON – The control of steam should have been on the steam supply line. Not based on the condensate back-up, due to the extreme differential pressure between the steam and process sides across the reboiler tube bundle.

XVI – Vacuum Tower Overhead Surface Condenser

A surface condenser that is fed by overhead vapor from a vacuum tower is in fouling service, as is a surface condenser fed by vapor from steam ejectors on a vacuum tower overhead. It is therefore necessary to be able to disassemble any surface condenser in vacuum tower service, so that the surface condenser can be cleaned.

It is difficult to clean the shell side of a surface condenser with fixed tube-sheets. Hence one can expect a decline in heat transfer as the condenser fouls, coupled with increasingly poor vacuum.

LESSON – A surface condenser with fixed tube sheets in vacuum tower service is a "throw-away" item. Do not accept a surface condenser with fixed tube sheets in new vacuum unit designs or as replacement for existing surface condensers.

The seal legs which drain the condensed vapor from the surface condenser to the seal drum in a vacuum tower overhead system may become partially plugged.

If it is not possible to clean out the seal legs of the surface condenser while the Vacuum tower is in service then the condensate will gradually back-up into the condenser shell covering more and more of the condenser tubes, thus reducing the surface area for heat transfer, and causing a steady decline in condenser efficiency, increasingly poor vacuum in the vacuum tower, and possibly causing the steam ejectors to surge.

Check for condensate back-up in a surface condenser by measuring skin temperatures around the outside of the shell with a temperature gun, look for a sudden decrease in temperature that would correspond with the level of condensate backed up in the shell side.

Cleaning the seal legs on-line is by steaming out. There should be two valves for closing off the seal leg, one above the other at the upper end of the seal leg, with a stub valve connected to the side of the seal leg between them. A steam hose is temporarily connected to the stub valve in the seal leg to emit steam to the seal leg. Each of the valves above and below the stub valve are then closed one at a time whilst the other is left open. The objective is to force steam either up into the condenser shell so as to dislodge deposits at the top of the seal leg, or with upper valve closed and valve below the stub valve open to dislodge deposits in the lower part of the seal leg and also move sludge accumulating in the seal drum away from the end of the seal leg.

<u>LESSON</u> – In new designs, ensure that the two clean out valves are included at the top of the seal leg and that there is a side stub valve between them. If side stub valve, or clean out valves are missing on an existing surface condenser seal leg, plan on installing them at next unit shut down (Fig. XVI).

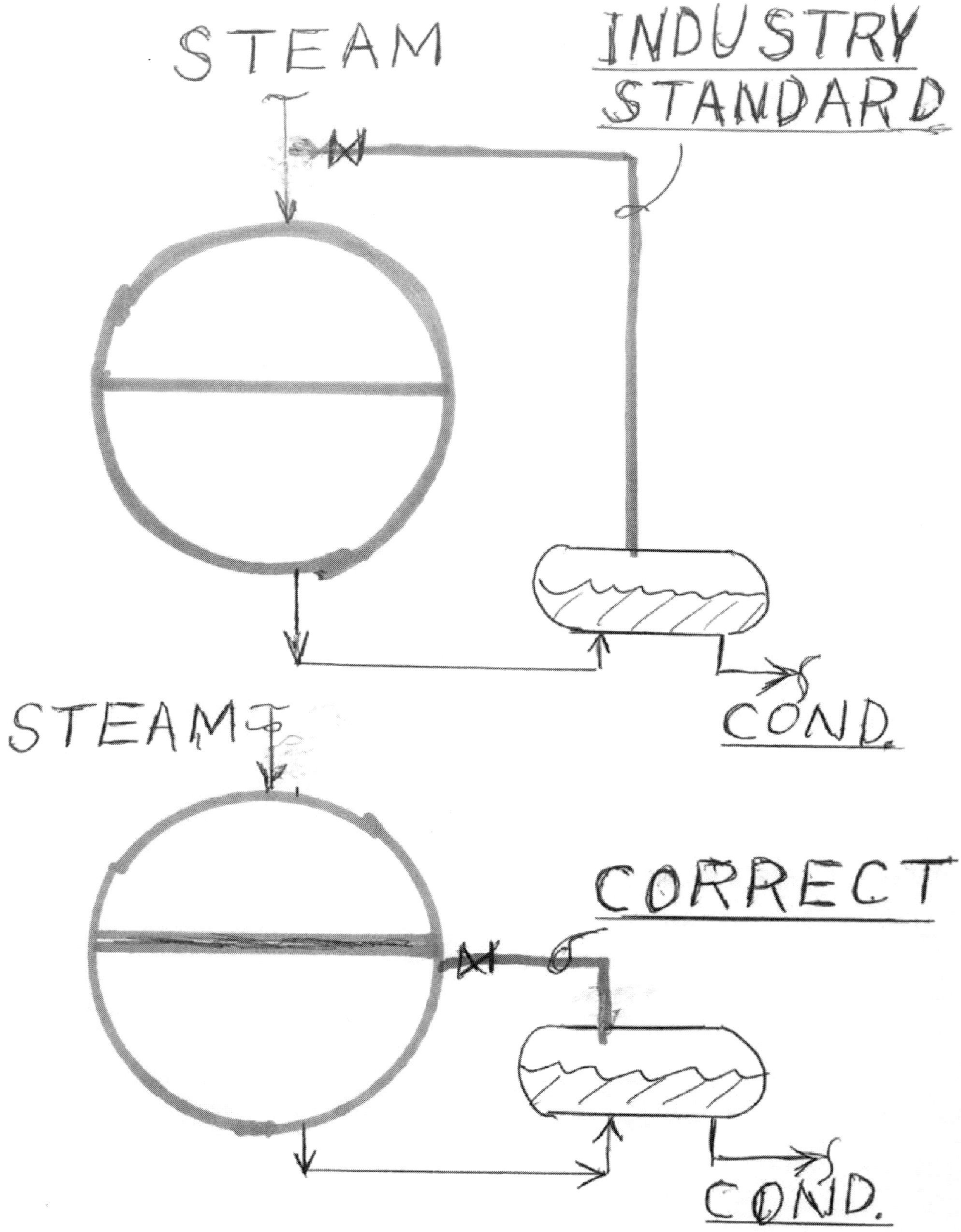

FIG. #1

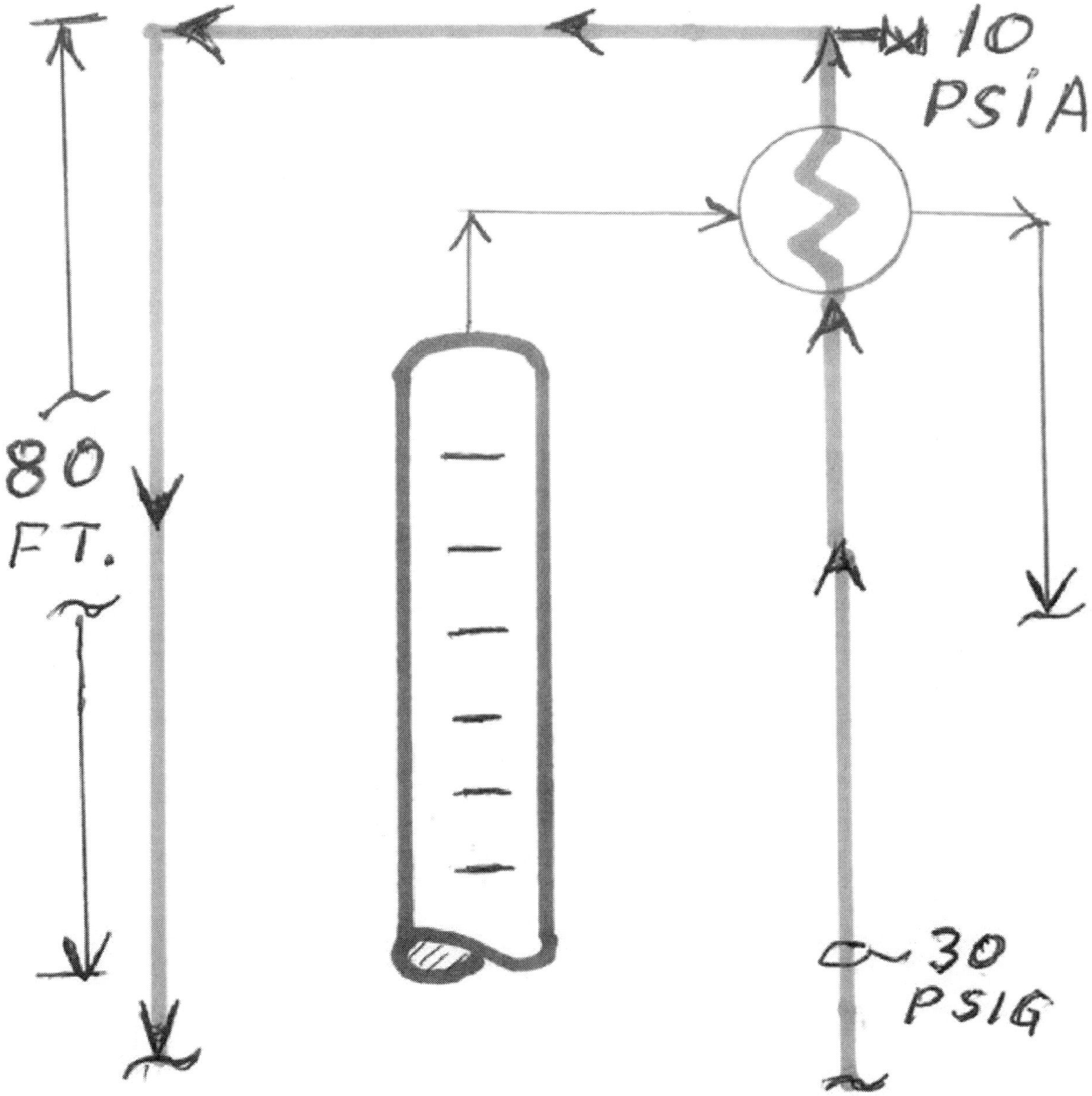

FIG. # II

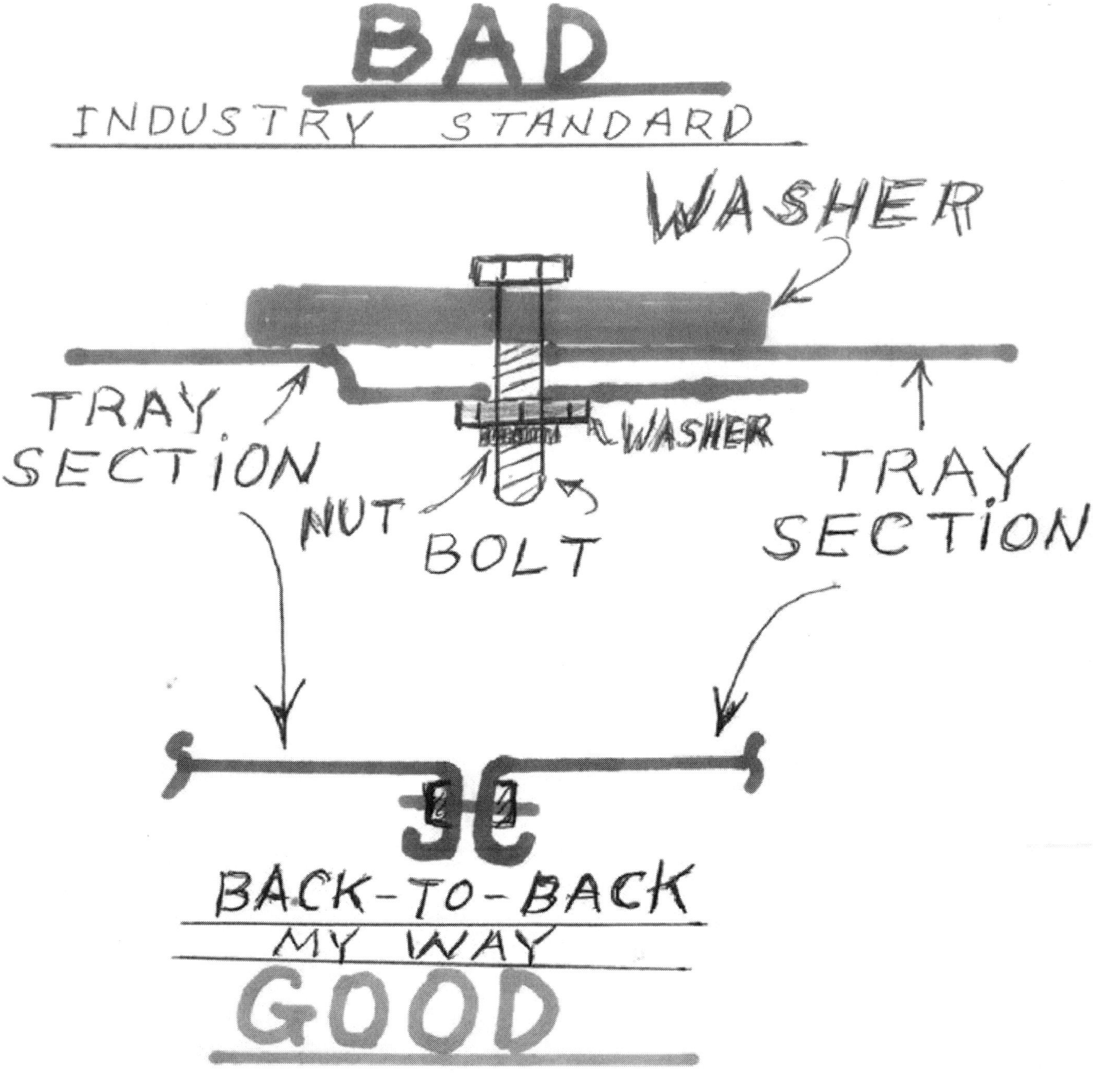
BAD
INDUSTRY STANDARD
WASHER
TRAY SECTION
WASHER
NUT
BOLT
TRAY SECTION
BACK-TO-BACK
MY WAY
GOOD

FIG #III

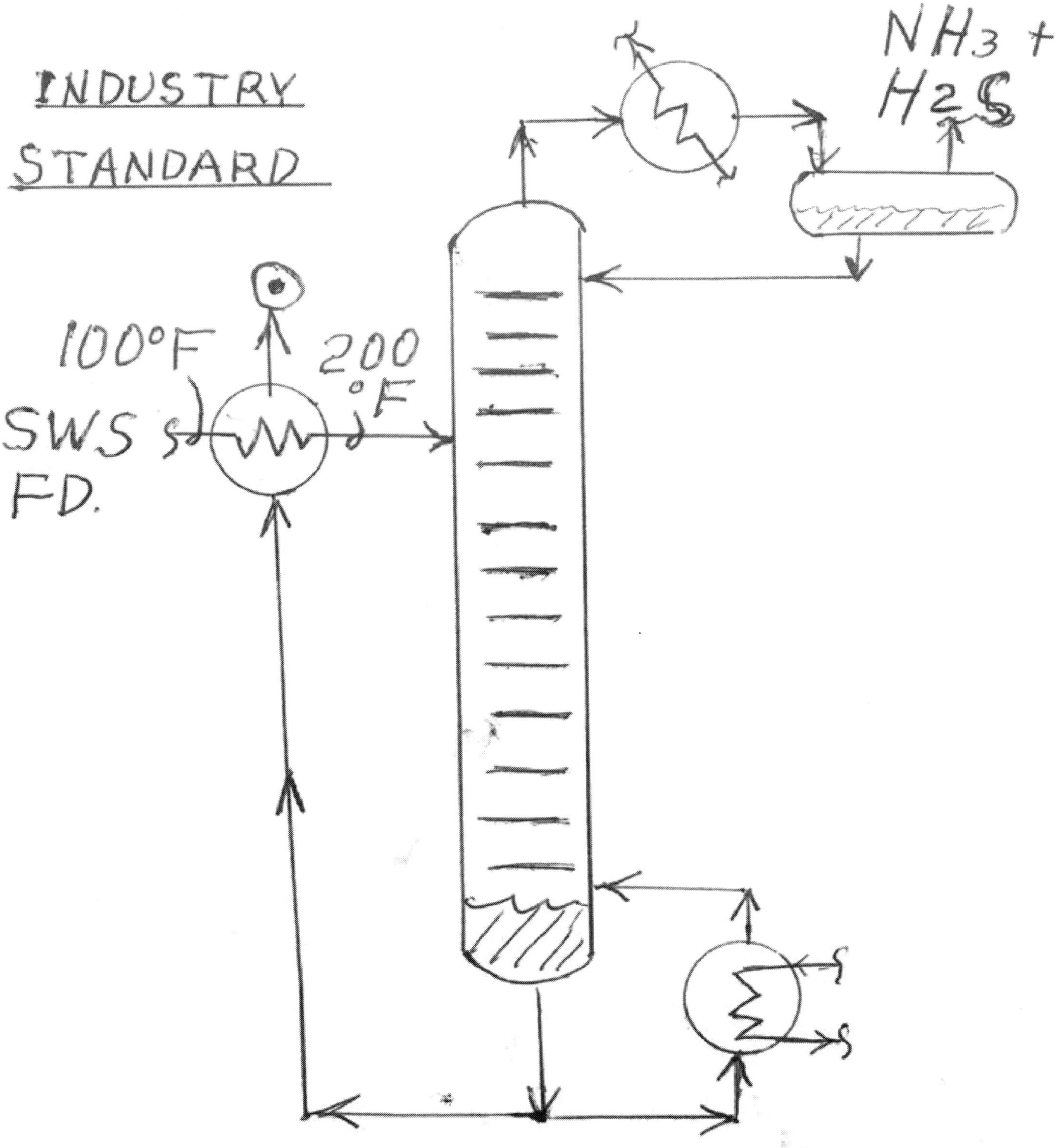

FIG. #IV-A

CORRECT DESIGN

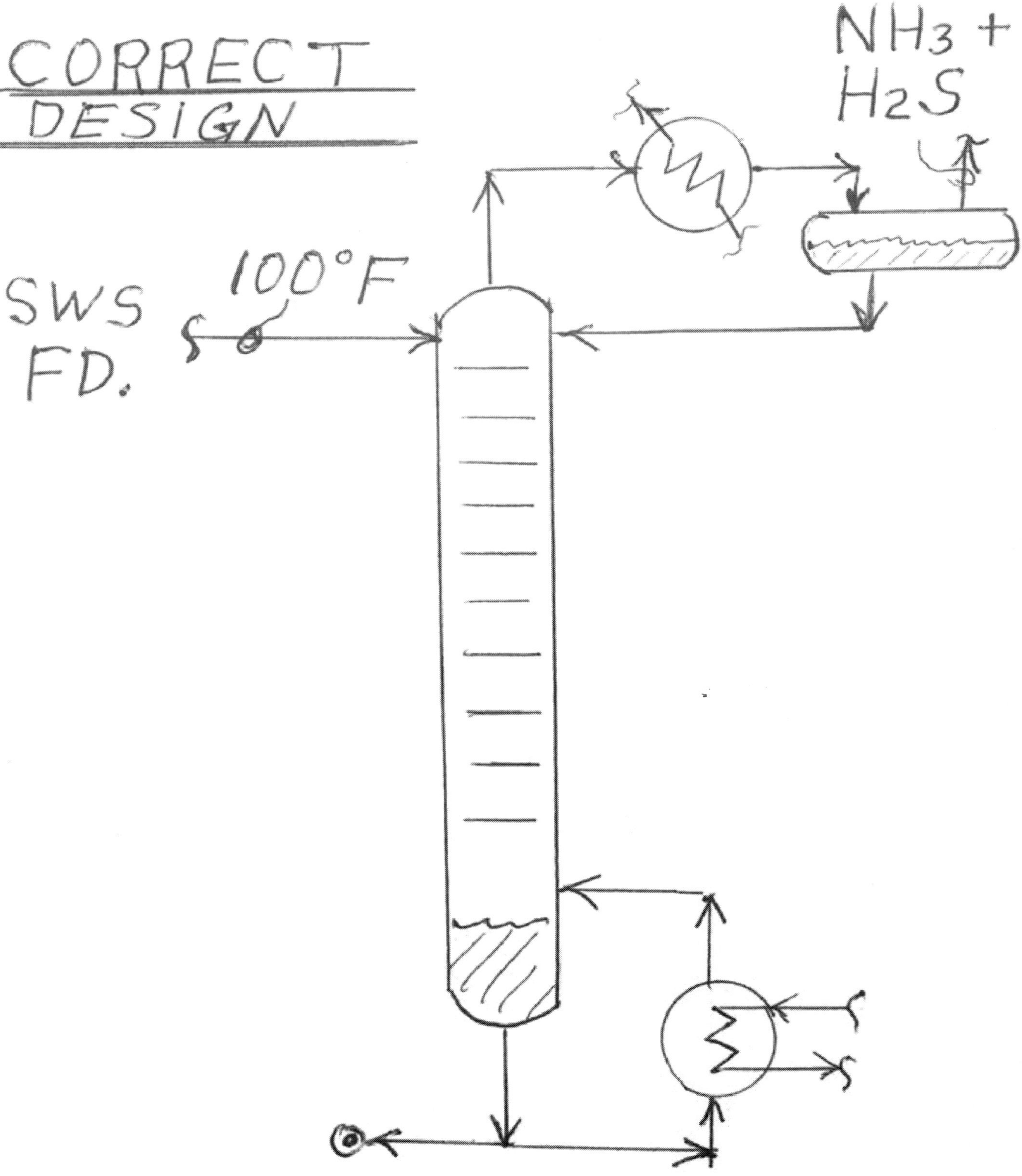

FIG. #IV-B

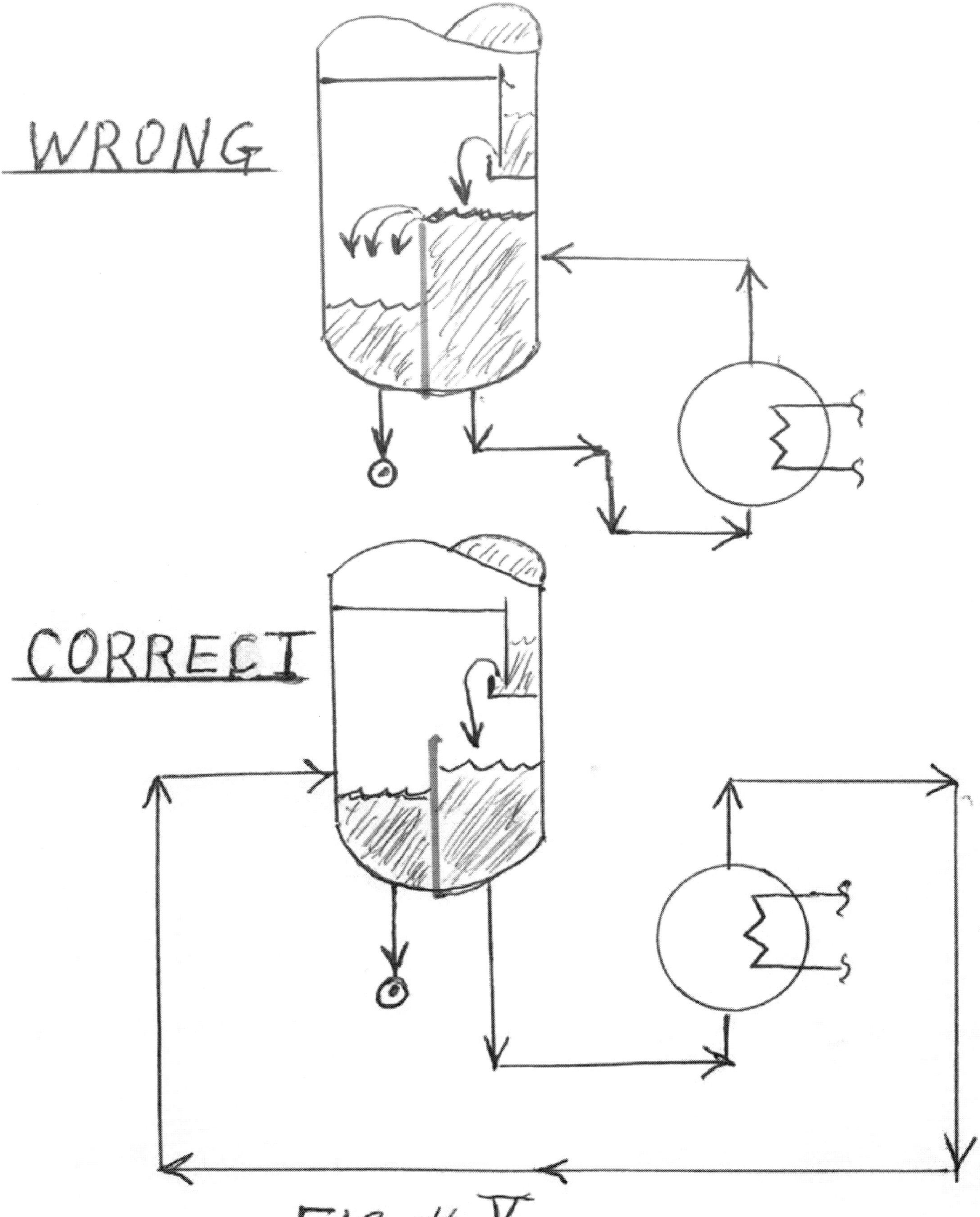

FIG. # V

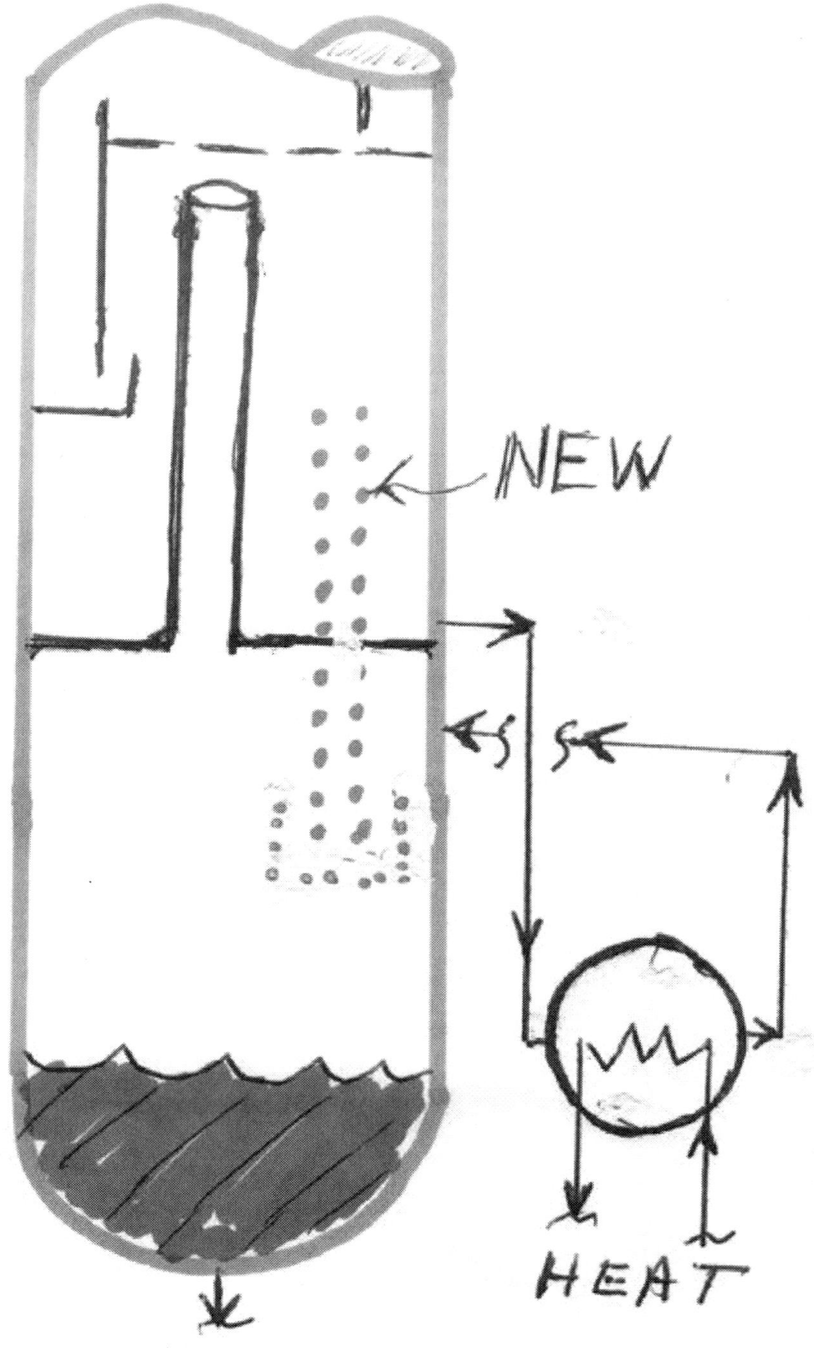

FIG. #VI

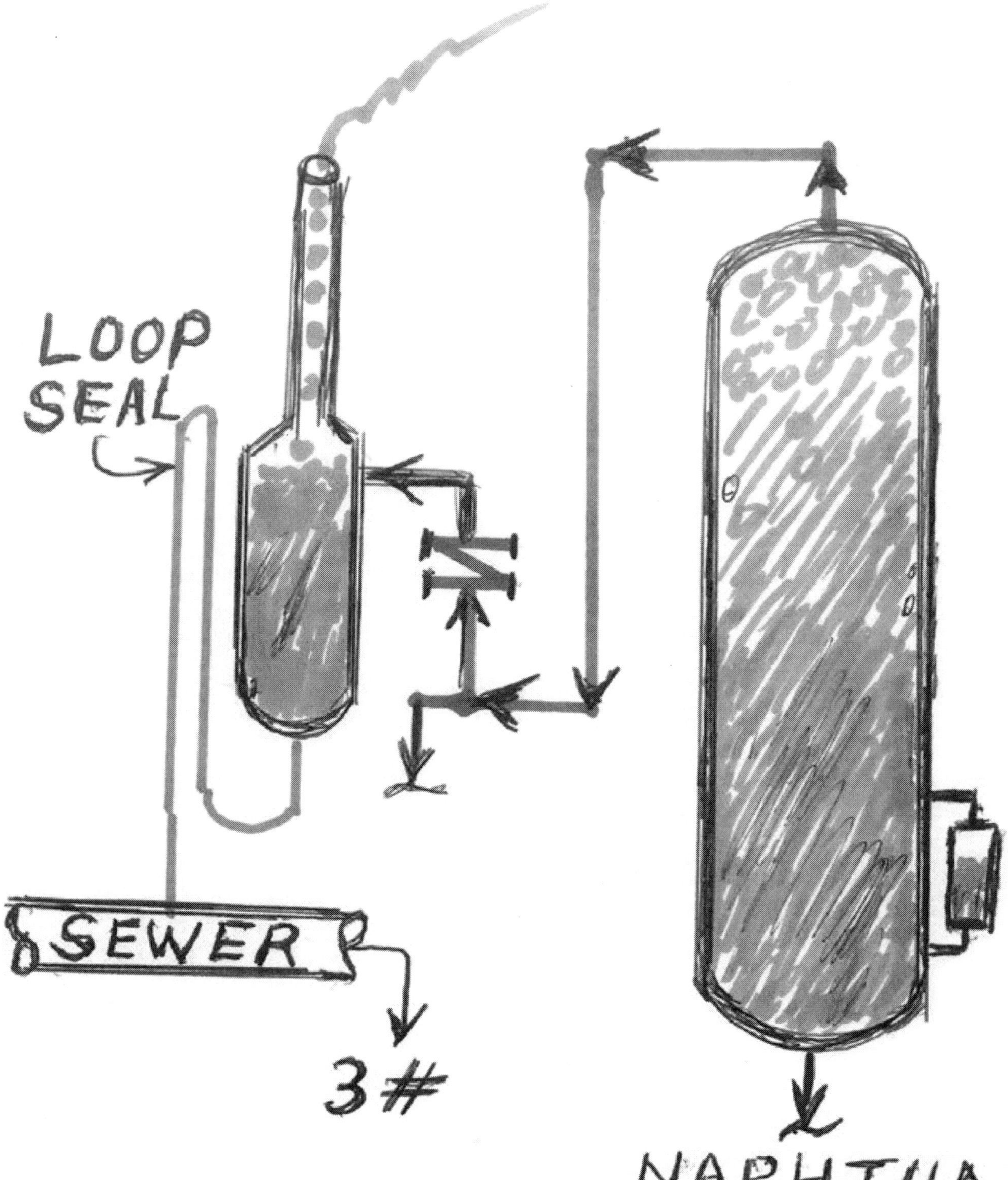

FIG. # VII

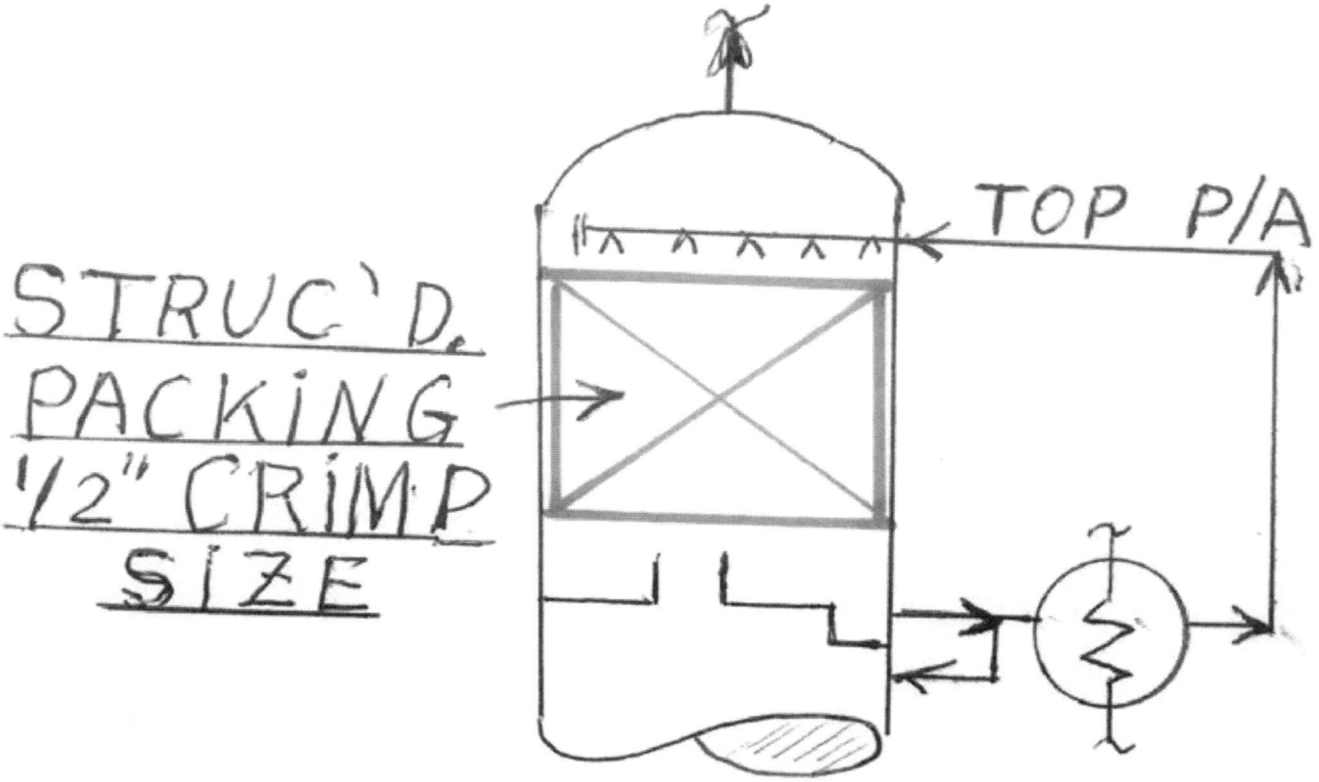

FIG. # VIII

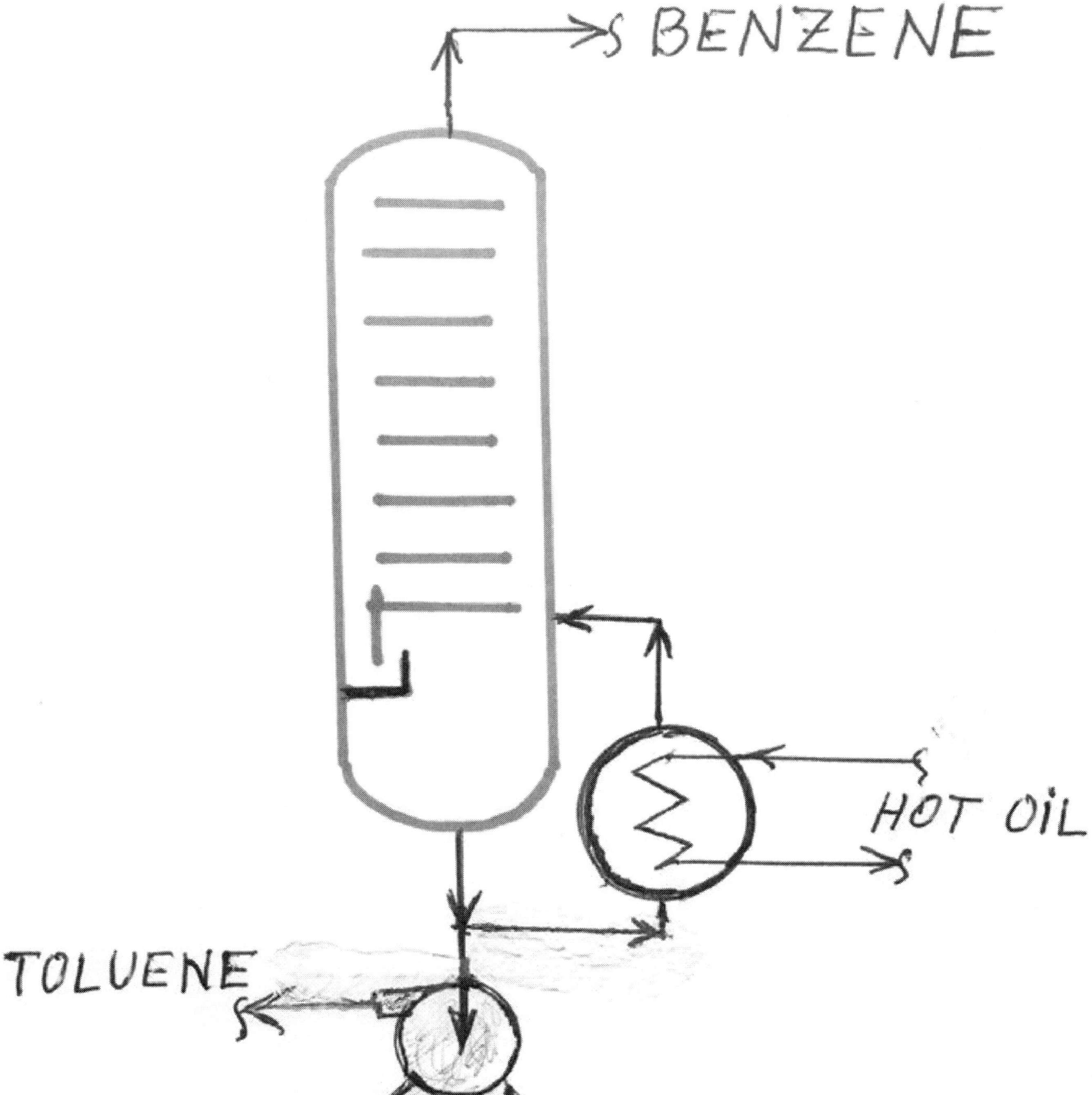

FIG # IX

SOUR WATER STRIPPER

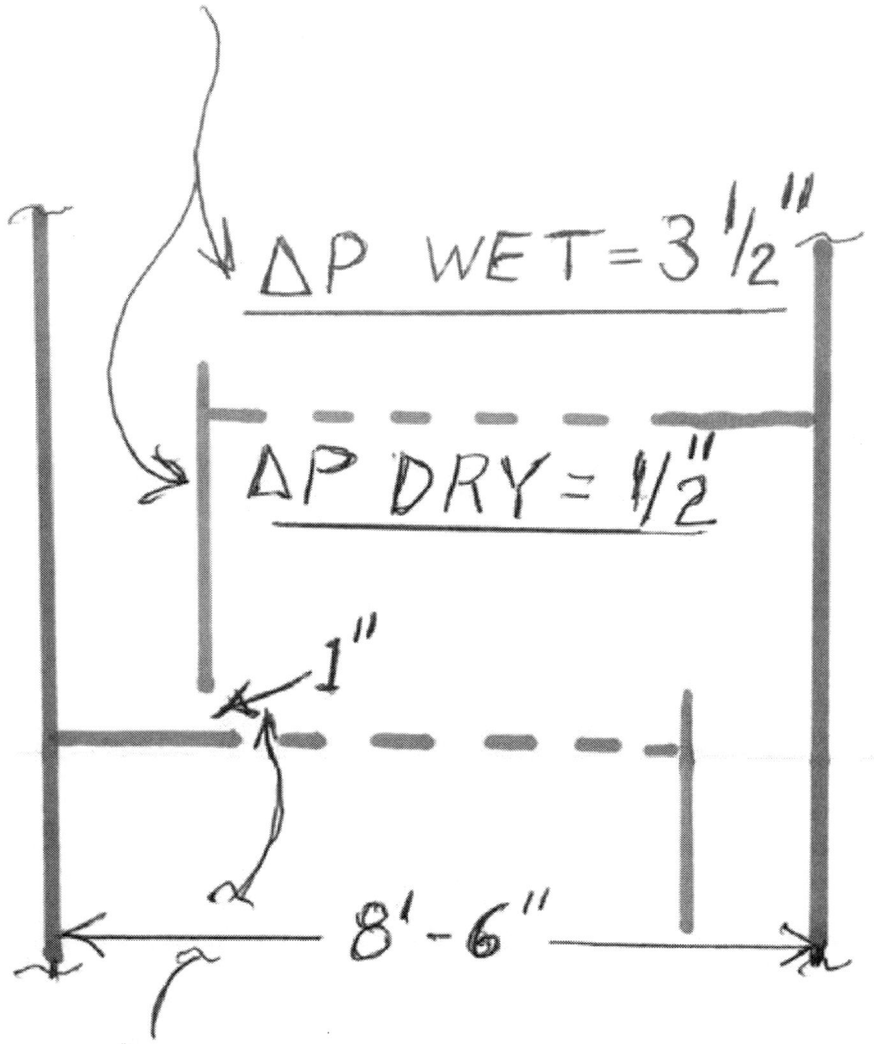

RESID STEAM STRIPPER

FIG. # X

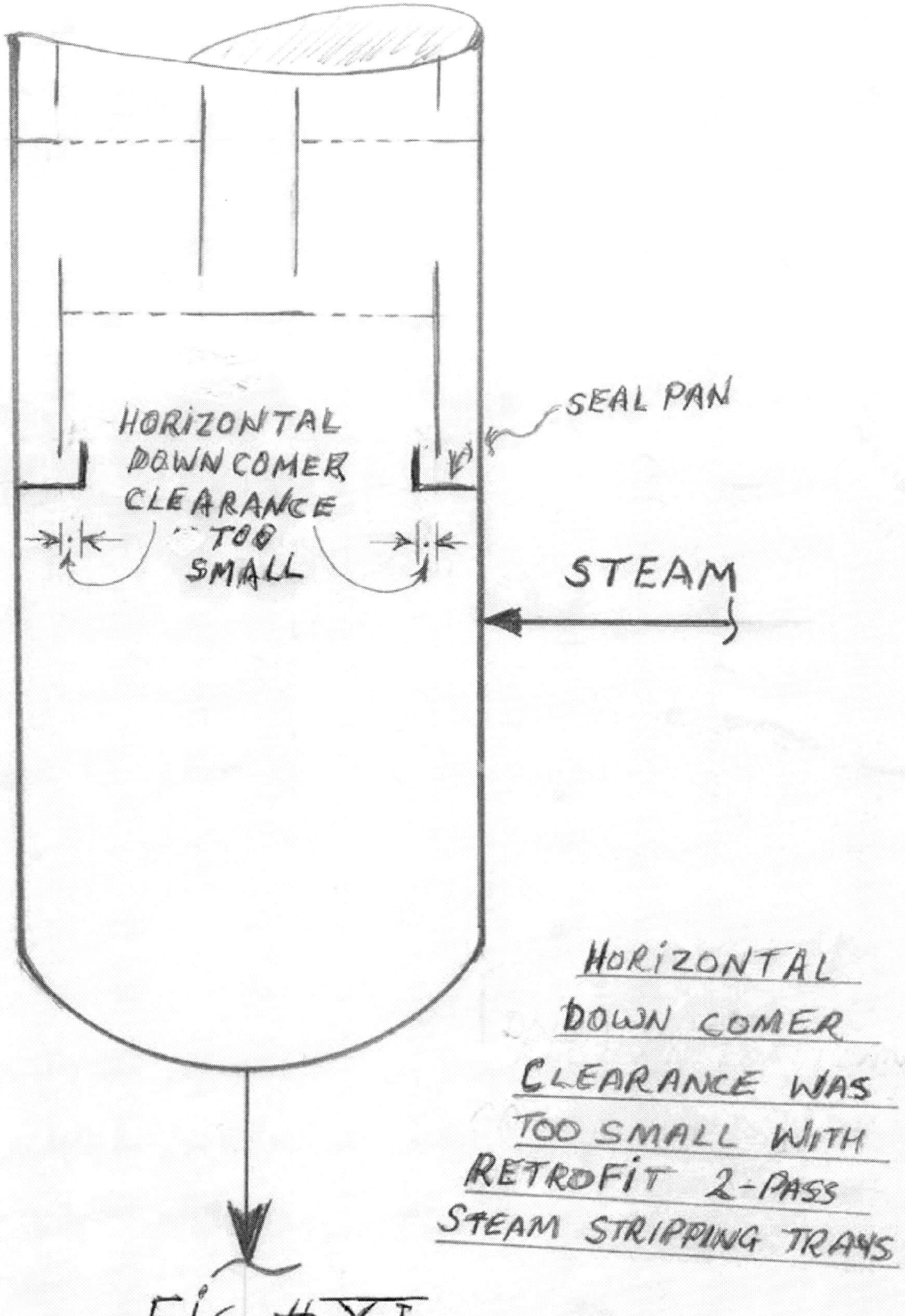
SEAL PAN
HORIZONTAL
DOWN COMER
CLEARANCE
TOO
SMALL
STEAM
HORIZONTAL
DOWN COMER
CLEARANCE WAS
TOO SMALL WITH
RETROFIT 2-PASS
STEAM STRIPPING TRAYS

FIG #XI

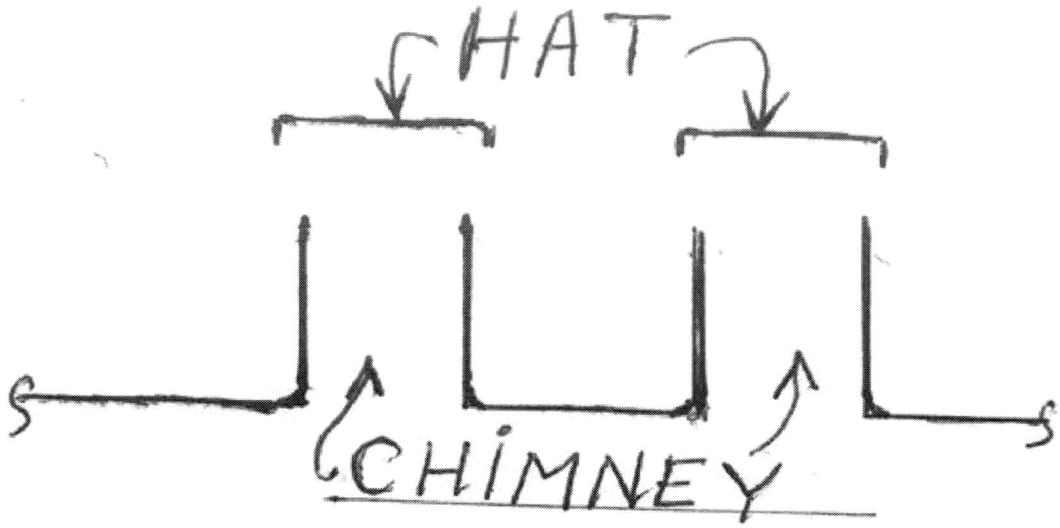

Fig # XII

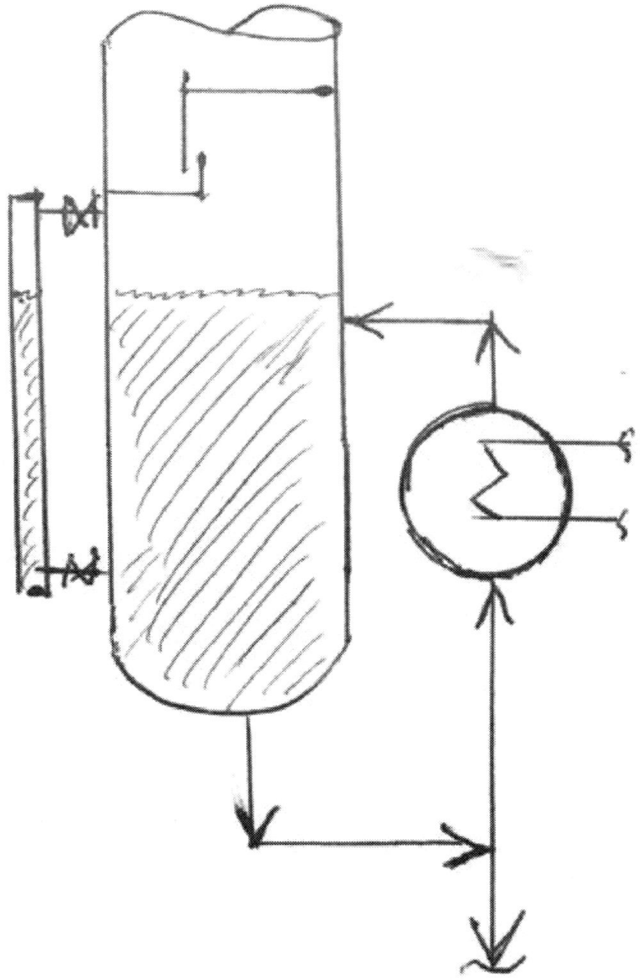

FIGURE # XIII-A

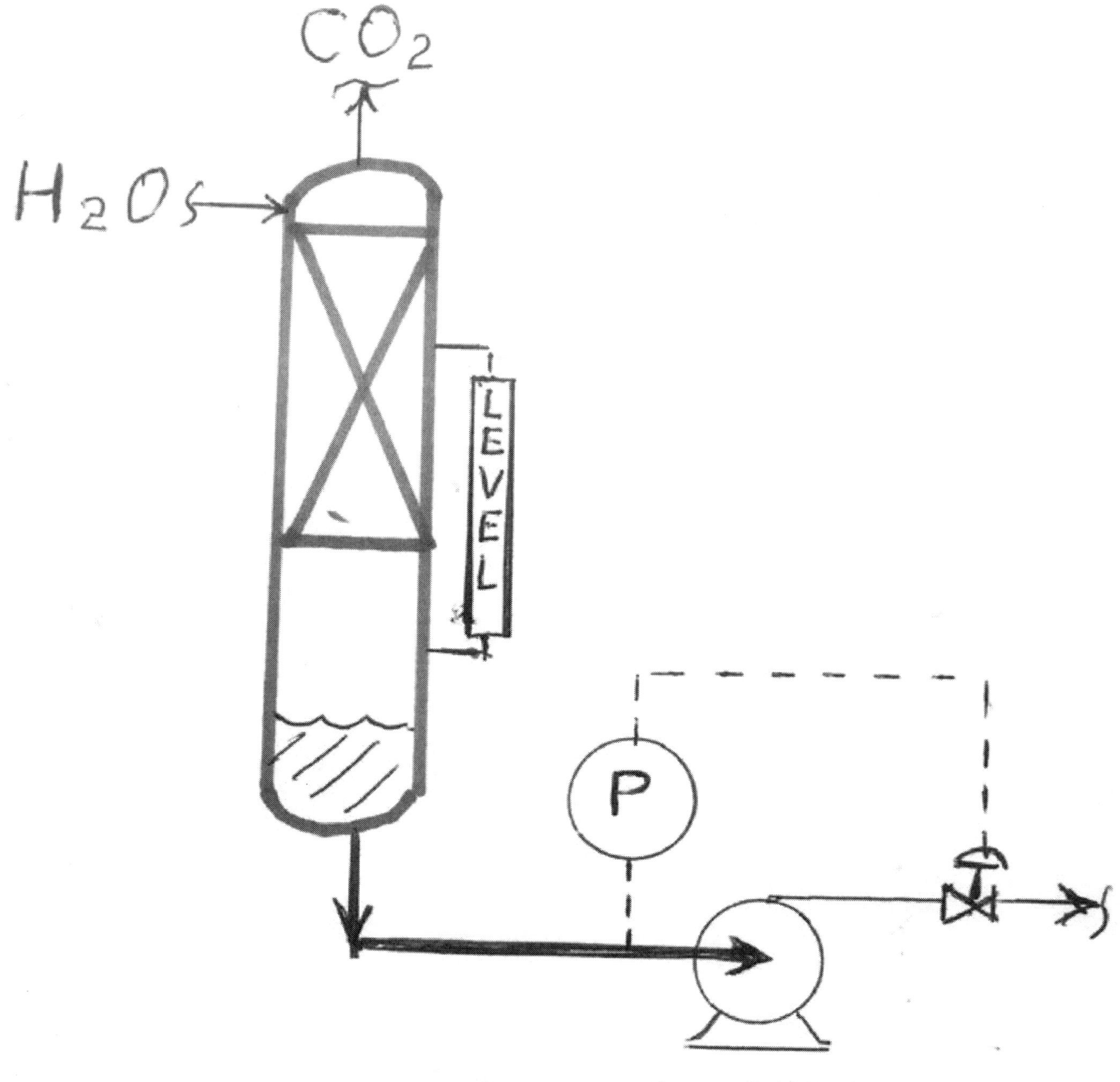

FIG. # XIII-B

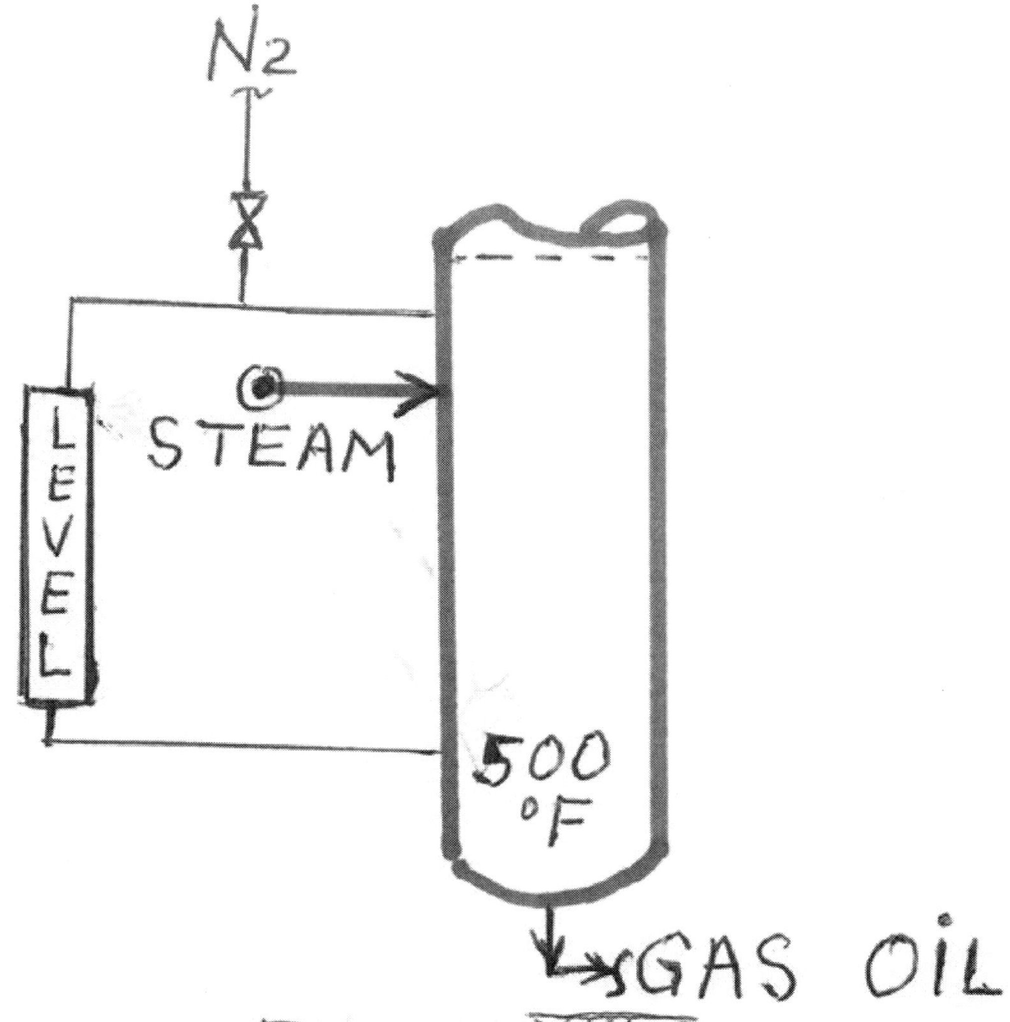

FIG. # XIIIC

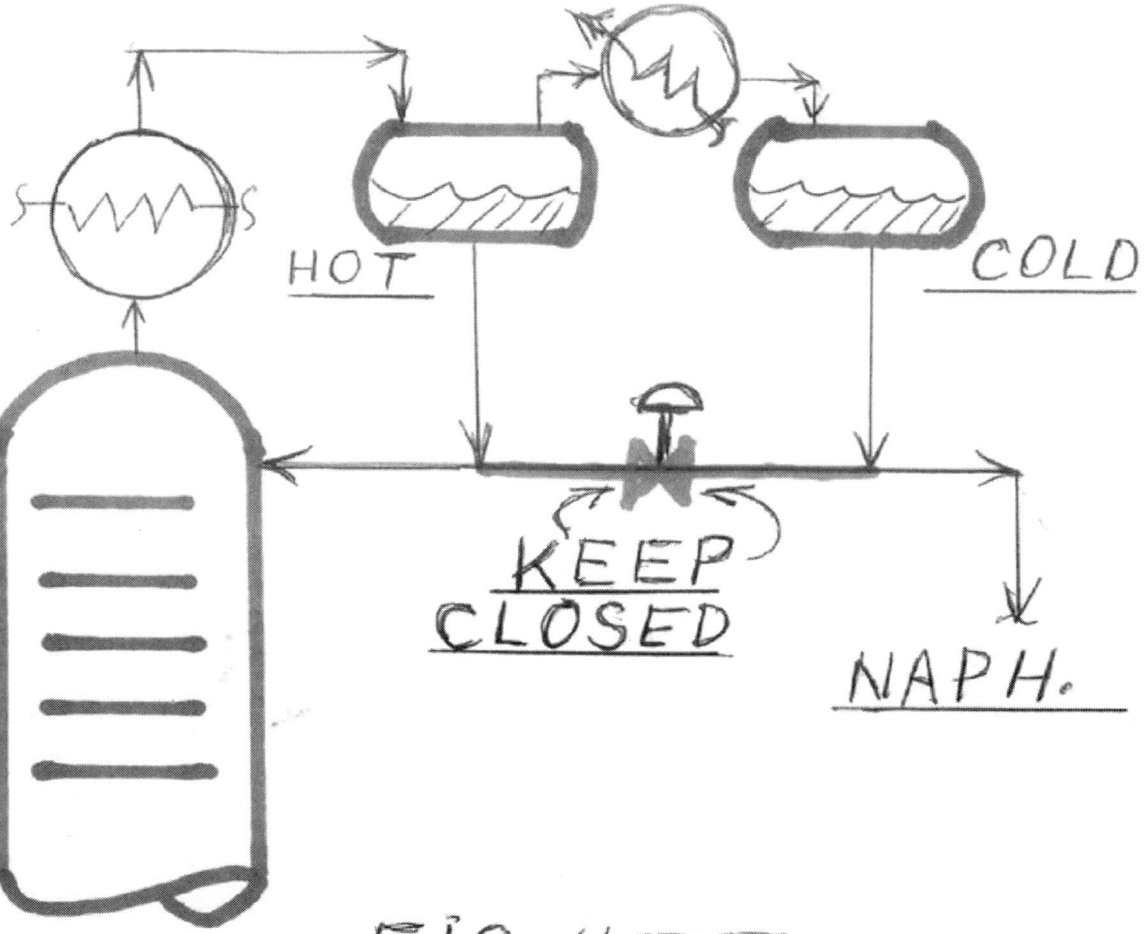

FIG # XIV

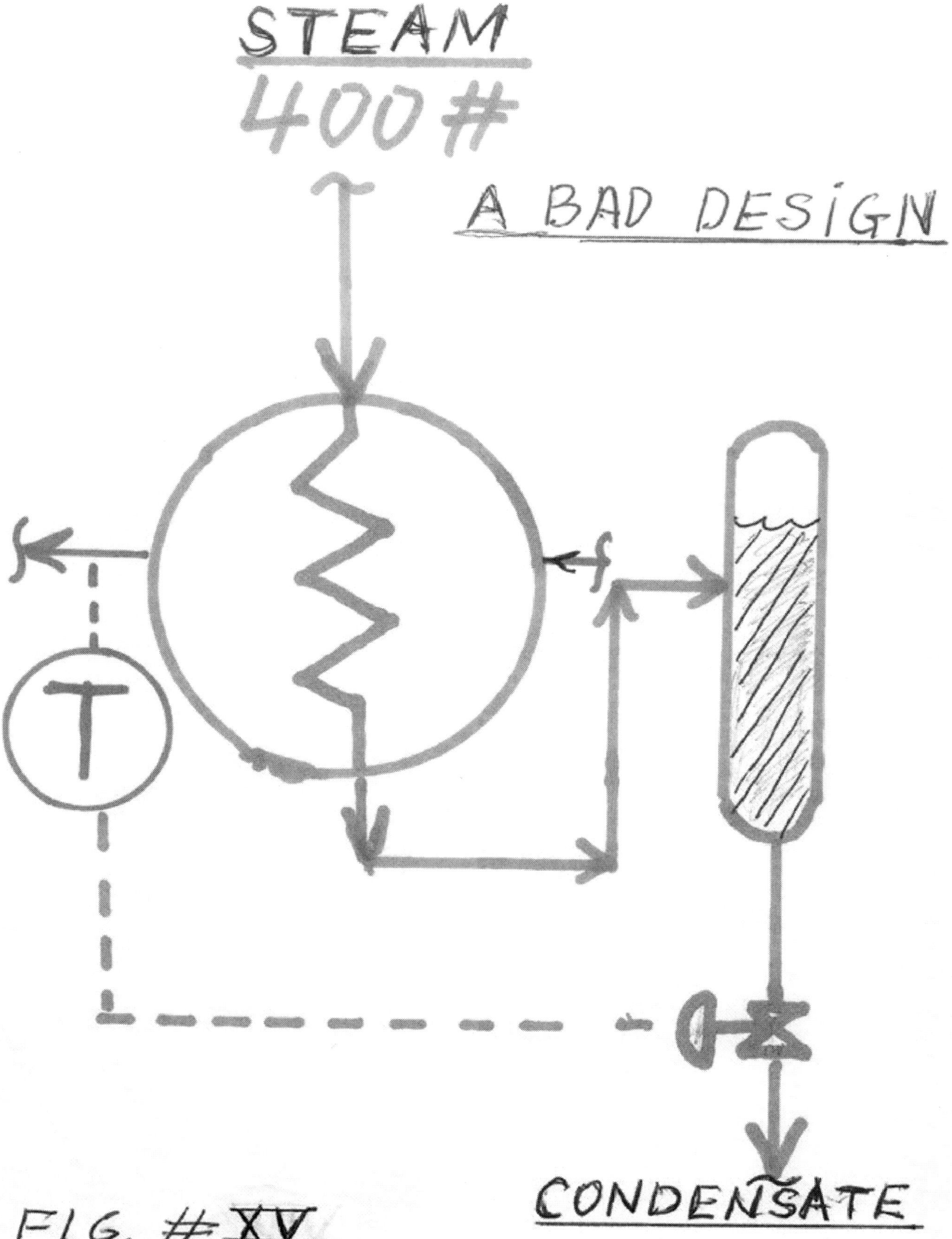

FIG. # XV

ON-LINE CLEANING
FOR SURFACE CONDENSER
SEAL LEG

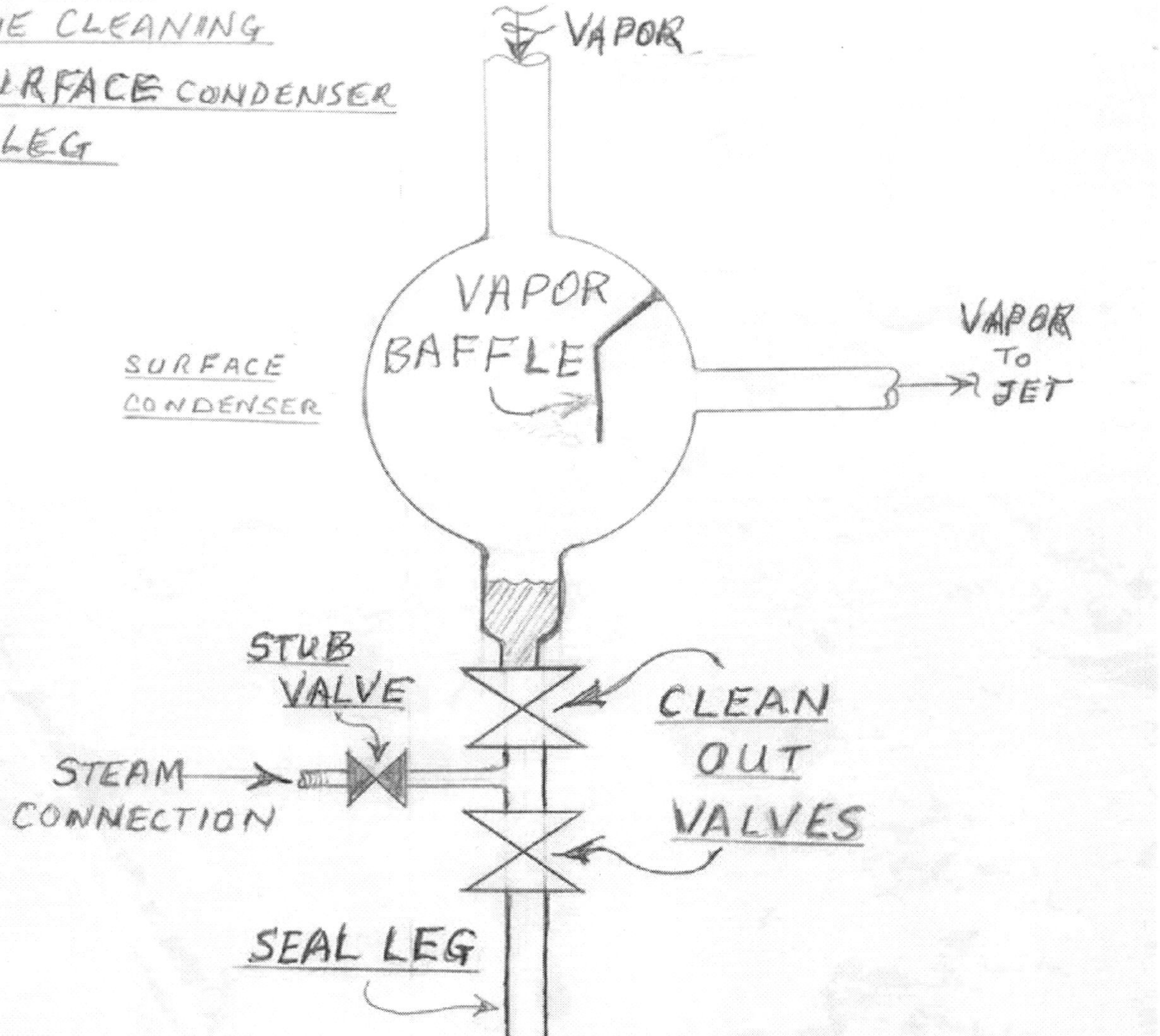

FIG. # XVI

551723 Changing Compositions Impacts Pressure Control

This paper describes a case study of a multi-pass distillation process with uncontrollable tower pressure.

A chlorosilane intermediate material is purified via distillation campaigns on a shared asset. Customer requirements for higher quality resulted in a need to process the material twice through the same towers to remove different impurities. During the second pass the tower pressure was uncontrollable and the plant has lived with this for years. Pressures would rapidly swing up and down causing dramatic shifts in column temperature profile and inventories inside the tower. This has resulted in low yield and product quality challenges. Several attempts were made to resolve the issue but with no success.

A thorough investigation revealed the composition of the feed during this second pass was not compatible with the installed pressure control scheme. The tower runs at vacuum pressure for this product and the control scheme requires the presence of a higher volatility species to hold tower pressure. The volatile components are removed on the first pass leaving very little to hold pressure steady during the second. Once this was identified, the addition of a noncondensible was included in the pressure control scheme and this instantly resolved the problem. The process has been able to achieve record yields and reduce significant costs. No capital was needed to resolve this long term issue.

549435 Troubleshooting and Redesign of a Solvent Stripper in a Chemical Plant

A solvent recovery column in a chemical plant was revamped to increased capacity. During start-up at the revamped rates, the column experienced instability marked by pressure drop increases. In addition to an inability to reach desired production rates, the instability also caused off-specs products. The investigation found the problem to be downcomer flood due to foaming, a condition which worsened at high liquid rates. An antifoam was identified in a foaming study; the selected antifoam suppressed the foam layer allowing the plant to achieve design rates. The column tray deck was redesigned so the antifoam could be eliminated. The redesign is expected to provide a 17% additional capacity compared to the current design assuming an approach to flood of 75%.

551724 Common Refinery Distillation Design Errors Part 2: Tower Internals

Based on the authors refinery process troubleshooting experience, a list of a dozen common process design errors has been extracted from their seminar outline.

These errors range from the realm cause of the B.P. Texas City Raffinate Splitter Explosion, to a steam reboiler bottleneck that the authors corrected just a few weeks back in India on a Sour Water Stripper. Exactly why refiners persist in failing to identify the same errors, decade upon decade was a question posed by the mother of one of the authors.

551185 Recycling Nitric Acid Contaminated with Radionuclides

This presentation will cover the design, fabrication, and operation of a system to recycle nitric acid containing high levels of radionuclides. The challenge was to recycle as much nitric acid as possible while reducing waste. Just a few of the challenges for this project were high radiation levels, highly corrosive solutions, extremely tight water discharge requirements, and extremely limited system access after system start up. On top of these requirements, it was also desired to operate this as a continuous distillation that was only operated during one shift (restart system every morning), with very limited space requirements. It is also under severe time and budgetary constraints. We must also remember that this system must meet Nuclear QA/QC program requirements.

Diagnosis Key to Rectification of Premature Flood and Instability in a Debutanizer Tower

Henry Z. Kister, Fluor, Aliso Viejo, CA

Carlos Trompiz, Fluor, Houston, Texas

and

Brian Clancy-Jundt, Flint Hills Resources, Houston, Texas

Presented at the Kister Distillation Symposium, AIChE Spring Meeting, New Orleans, Louisiana, March 31 - April 4, 2019

SUMMARY & CONCLUSIONS

The Flint Hills Resources (FHR) Debutanizer tower in their Propane Dehydrogenation (PDH) Unit in Houston, Texas, experienced premature flood at high loadings. A sharp pressure drop rise was encountered when the tower internal vapor rate through the top section exceeded 13,000 to 15,000 lb/h. This increase in pressure drop was accompanied with an increase in the bottom C4 content. Instability and fluctuations of key process variables such as tower differential pressure and bottom flow rate occurred at all rates, intensifying when the tower flooded.

To diagnose and solve the problem, and to debottleneck the tower, Fluor, who was not involved in the original design but has extensive distillation expertise, was invited to join the FHR troubleshooting task force.

Plots of tower differential pressure (dP) versus vapor loads were prepared. These confirmed the flooding, and suggested that it was most likely occurring above the feed. Hydraulic calculations did not show proximity of any of the major flood limits. The percent jet flood for all trays was below 50%, downcomer liquid backups were very low, and so were downcomer inlet velocities. So no major flood limit should have been encountered.

Three gamma scans were performed at different rates. The latest of these, on February 17th, 2015, was performed at high vapor loads and high dP. The scans looked similar to each other. No flooding was observed in any of the gamma scans.

Closely reviewing the hydraulics revealed extremely low head losses at the downcomer exits of the trays above the feed, about 0.11 inches of liquid. This raised the suspicion of a possible loss of downcomer seal. Applying the Fluor downcomer unsealing model confirmed that vapor was likely to break the seal on these downcomers, and to generate downcomer unsealing flood. Unsealing of downcomers may have also occurred below the feed, but our model showed that there it is most likely that the unsealing only led to tray efficiency reduction but not to flooding.

The downcomer unsealing flood was unable to explain the fluctuations, as they also occurred when the tower was not flooded. Further investigation traced the root cause of these fluctuations to the unvented vapor space above the baffled bottoms draw compartment. Trapped unvented gas generated backpressure on the reboiler compartment liquid with intermittent geysering and fluctuations, as detailed in this paper.

To solve the problem, and with economics favoring reducing tray spacing to increase the number of trays, the tower was retrayed. Based on the above diagnosis, the new trays had effectively lower downcomer clearances. A gooseneck vent was added to vent the bottom draw compartment, and the downcomer from the bottom tray was shortened, reducing the submergence of the bottom downcomer in the sump liquid.

The tower is now operated with better separation than ever at 10% higher loads than previously with no flooding, no fluctuations, and at low pressure drops, attesting to the value of good diagnosis for economically solving tower problems.

OPERATING EXPERIENCE

Feed to the PDH debutanizer is about 50% mole C4 hydrocarbons which end up as an all-liquid overhead product and about 50% mole heavies, mostly C6 hydrocarbons, that end up in the bottoms. The C4 content of the bottom is below 1% mole. The tower operates at 80 psia at the top and a reflux to distillate ratio of about 3-4:1.

Figure 1 shows key tray and downcomer dimensions. The tower is 32" in diameter and contained large-opening, fouling-resistant valve trays. Above the feed the trays were conventional with 14% slot area (as a fraction of the active area), spaced 21" apart, and had straight 4.5" wide downcomers. Weir heights were 1.75" and downcomer clearances were 1.38". Below the feed the trays were at 18" spacing, with stepped truncated downcomers. The downcomers in this section terminated 4" above the tray floors, with 32% of the downcomer floor perforated to allow liquid descent. Right under the perforated downcomer bottom there were fixed valves. The fixed valve caps prevented downcomer liquid from weeping through the valves. The slot area of the valves on the trays below the feed was 15% of the tray active areas.

The tower experienced instability and high differential pressures (dP) at high loadings. Figure 2 is a historical record, showing a rapid rise in tower pressure drop at a tower vapor rate of about 13,000 to 15,000 lb/h. The vapor rates plotted on the x-axis were calculated from the sum of the product and reflux flow meters and are not corrected for reflux subcooling, which is likely to raise the vapor rate shown by around 10%. The blue points that show a rapid rise in pressure drop at about 8,000 lb/h and the green points showing another rapid rise at 11,000 lb/h are due to fouling episodes and are not addressed in this investigation. The focus here is on the pressure drop rise between 13,000 and 15,000 lb/h. This rapid rise in pressure drop conclusively indicates flooding.

There had also been fluctuations of key process variables such as bottom level, tower differential pressure and bottom flow rate. These fluctuations occurred at all rates, but intensified when the tower flooded.

TROUBLESHOOTING

The active areas of the trays in the tower were gamma scanned on three occasions: March 28th, 2012, February 12th, 2015, and February 17th 2015. The vapor traffic near the top of the tower (unadjusted for subcooling) corresponding to these scan periods were around 9,500 lb/h, 12,400 lb/h, and 14,700 lb/h, respectively. The pressure drop ranges were 2.1-2.2 psi, 2.8-2.9 psi, and 3.6-4.2 psi, respectively. The March 28, 2012 scans were performed at the lowest vapor loads, and the February 17th 2015 scans at the highest, about 30-50% higher than the loads on March 28th 2012.

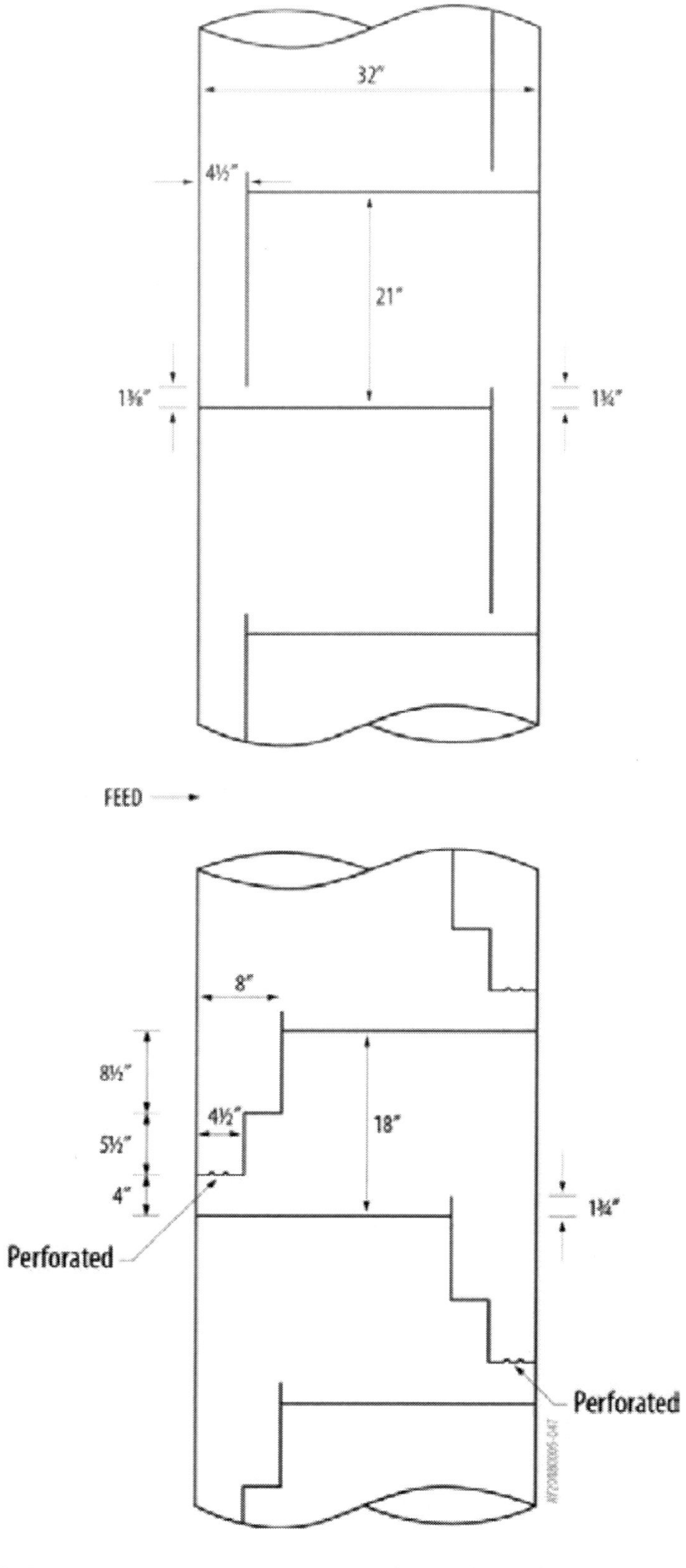

Figure 1. Key tray and downcomer dimensions for the debutanizer

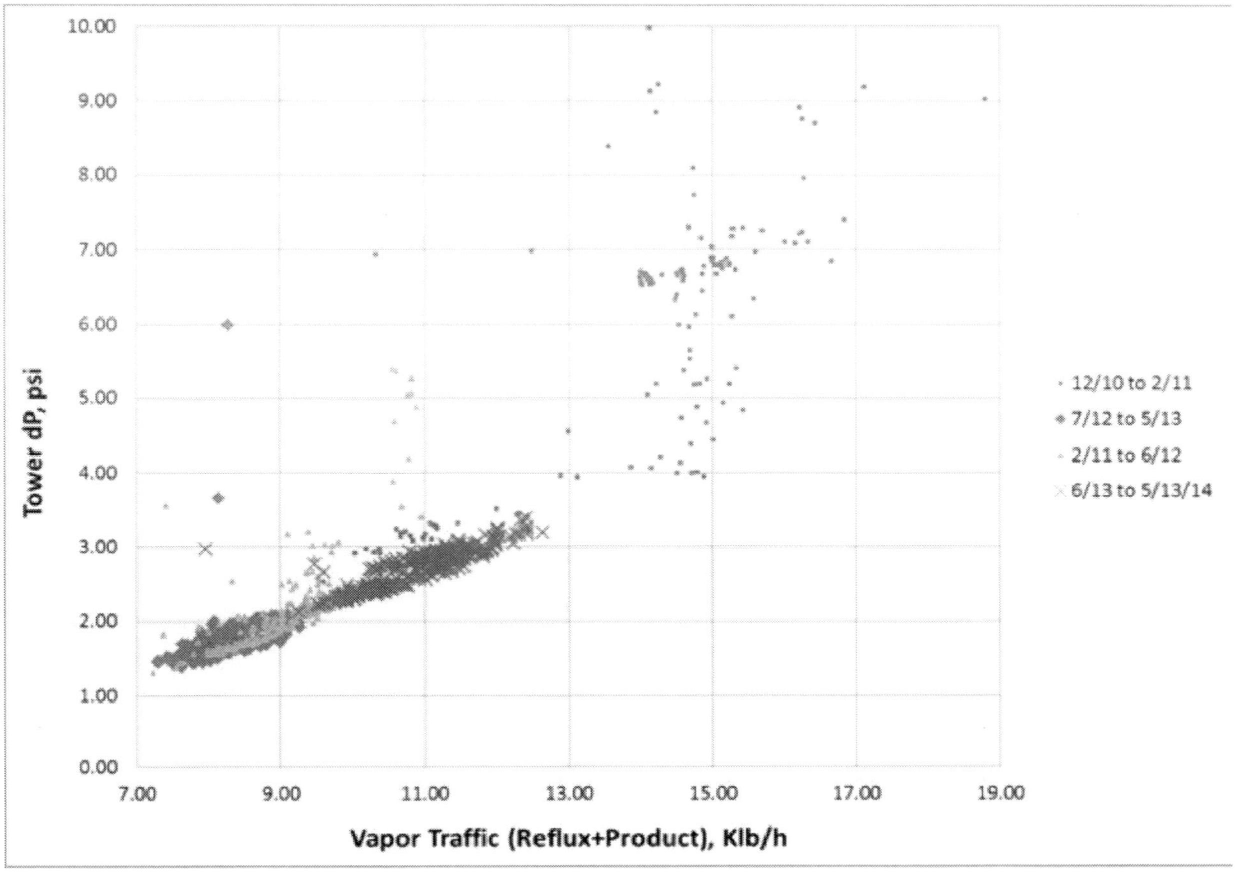

Figure 2. Flood Curves of the Debutanizer

Figure 3 shows the scans for the upper trays and the lower trays on all three dates. The intermediate trays, including the feed tray, are not shown but their scans showed identical trends and patterns as the trays shown in Figure 3. Figure 3 shows that despite the large differences in vapor loads and pressure drops there was very little difference between the three scans. None of the scans showed any flooding. In both the 2015 scans, the froth heights ranged from 38% to 51% of the tray spacings. If there was any entrainment from the trays, it was light.

During the February 17th scans, the reflux rate was intentionally raised to the point where flooding would be indicated by the dP transmitter, the tower was closely watched, and the following observations were recorded:

- Reflux set point was increased causing the dP to climb to 4 psi. The reflux set point was stabilized for approximately 15 minutes before the scan was started.
- Tower dP varied between 3.6-4.4 psi while the scan was being performed. Under normal conditions tower's dP fluctuation varies between 2.8-3.2 psi. This is nearly a 100% increase in variation.
- The debutanizer bottoms level fluctuated between 55%-85% full while the scan was being performed.

- Debutanizer bottoms C4 concentration increased from 1% to 2.6% mole nearly 15 minutes after the scan was complete and before the reflux was lowered back to normal.

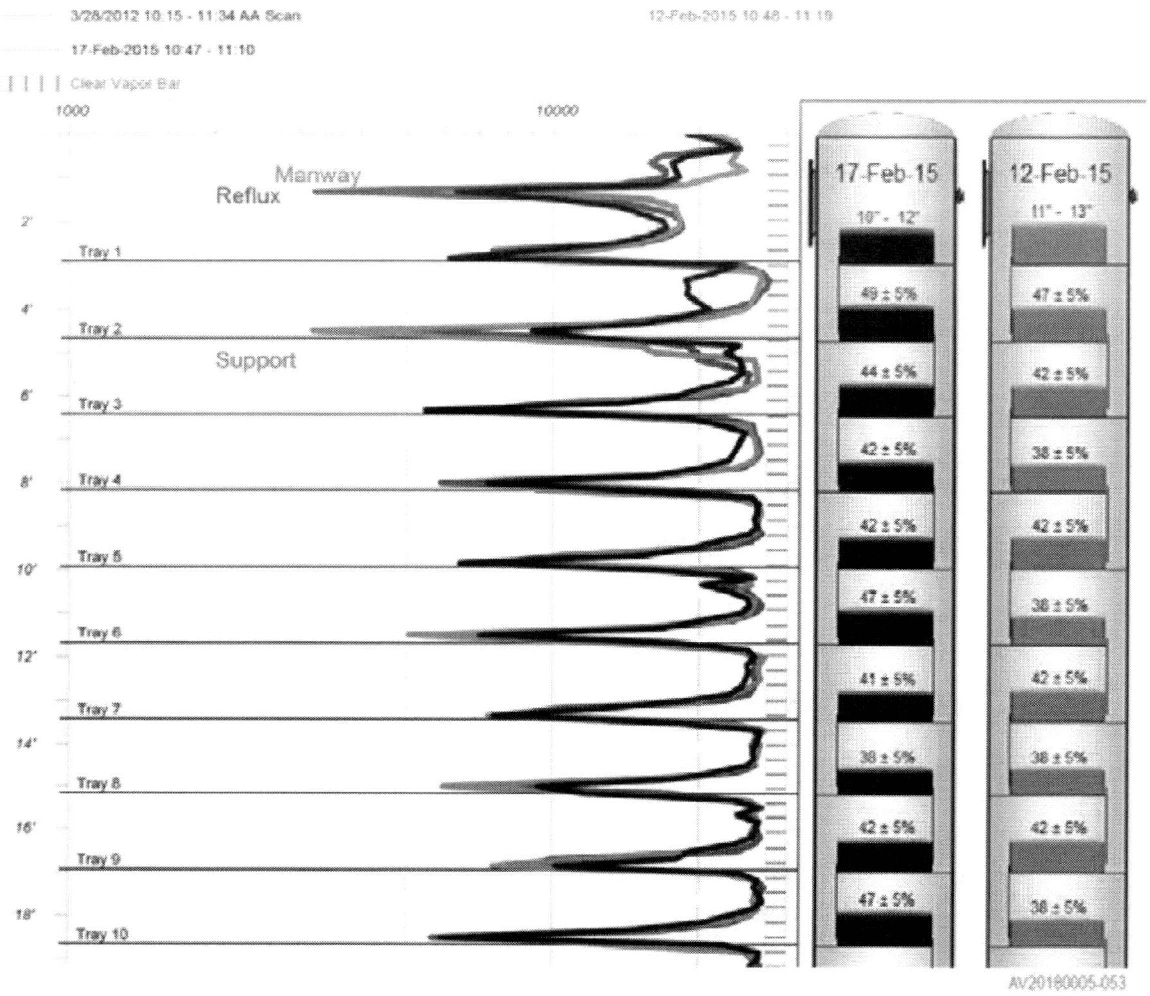

(a)

Figure 3. Gamma Scans of the Debutanizer, Flooded and Unflooded Conditions. No Flood is Indicated.

(a) Upper 10 trays, (b) Lowest 12 Trays, Including Bottom Sump

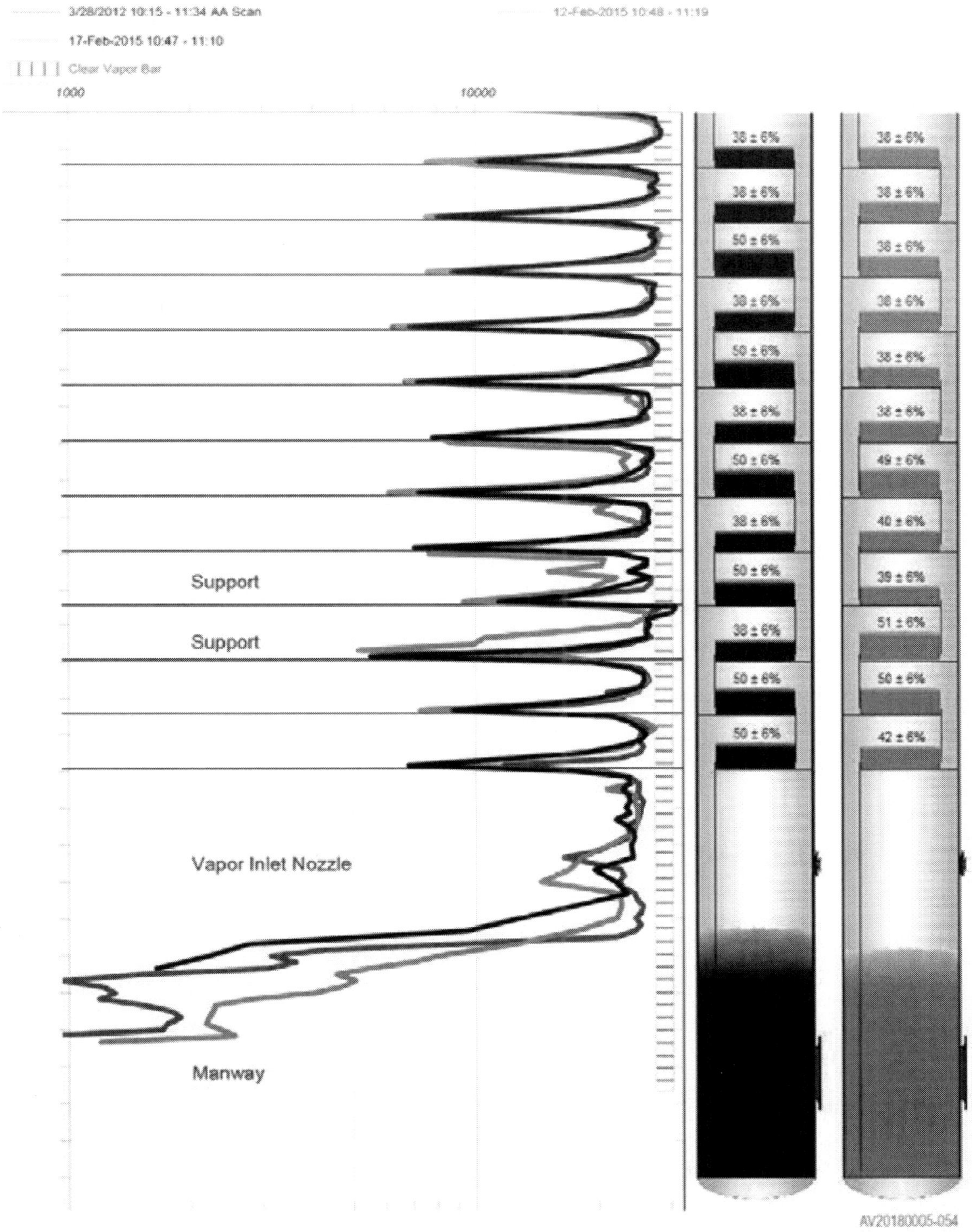

3/28/2012 10:15 - 11:34 AA Scan
12-Feb-2015 10:48 - 11:19
17-Feb-2015 10:47 - 11:10
Clear Vapor Bar
1000
10000
Support
Support
Vapor Inlet Nozzle
Manway
38 ± 6%
38 ± 6%
50 ± 6%
38 ± 6%
50 ± 6%
38 ± 6%
50 ± 6%
38 ± 6%
50 ± 6%
38 ± 6%
50 ± 6%
50 ± 6%
38 ± 6%
38 ± 6%
38 ± 6%
38 ± 6%
38 ± 6%
38 ± 6%
49 ± 6%
40 ± 6%
39 ± 6%
51 ± 6%
50 ± 6%
42 ± 6%
AV20180005-054

(b)

Figure 4 is a plot showing the pressure drop against vapor load near the top of the tower (reflux plus distillate) for the entire 24 hours period from 0 hour to 24 hour on February 17th. In this diagram (as distinct from Figure 2), the vapor load has been corrected for the reflux subcooling. The steep rise in dP in Figure 4 confirms that according to the dP, flooding was occurring during the scanning period on February 17th. Similar to Figure 2, Figure 4 shows a good correlation between the steep rise in dP, which indicates flood, and the vapor traffic near the top of the tower.

Figure 4 DC4 dP versus corrected top vapor flow rate

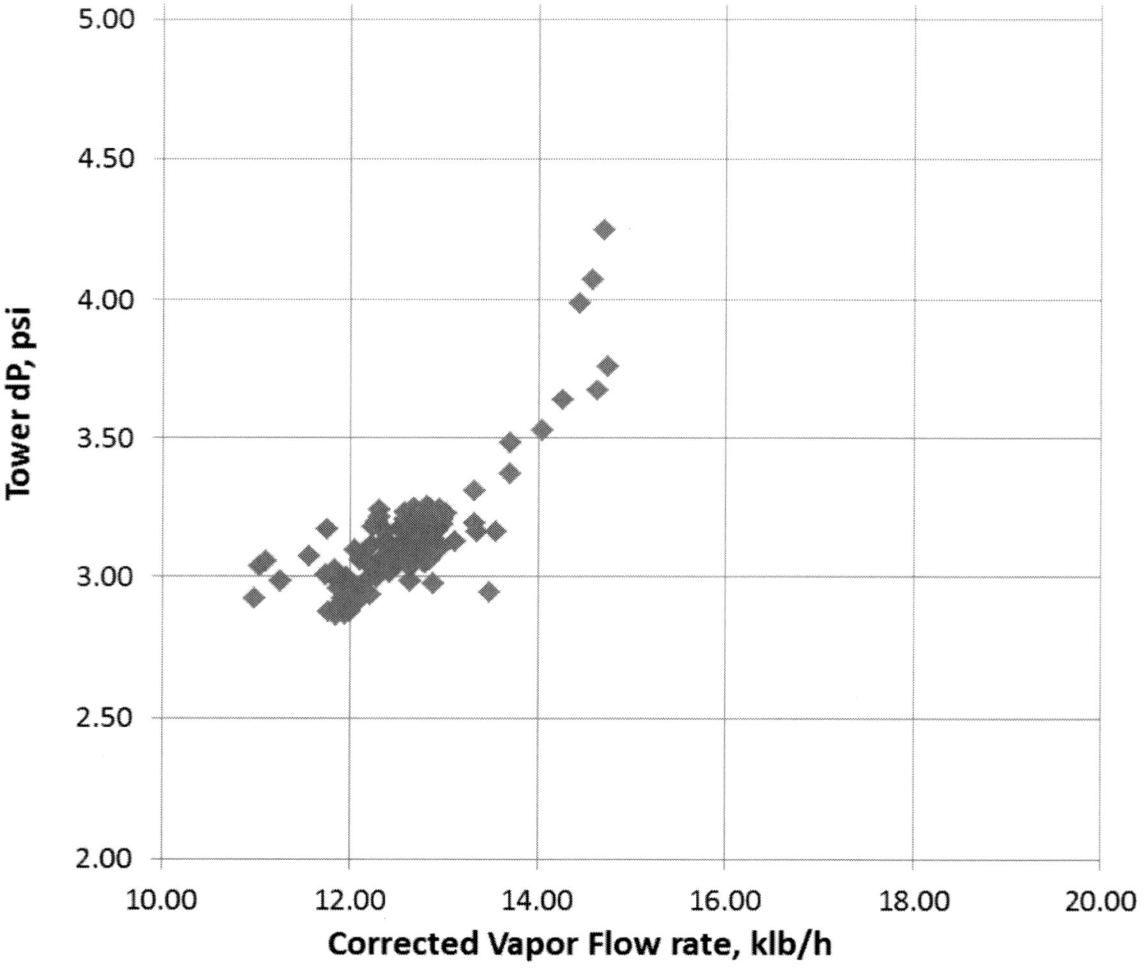

Figure 5 plots the tower pressure drop against the steam flow rate. This diagram shows a much weaker correlation between the dP and the steam rates than the correlation with the top vapor rates in Figure 4. There were a few periods on February 17th where the steam rate was high but the dP did not take off. This provides non-conclusive evidence suggesting that the flooding is likely to occur above the feed.

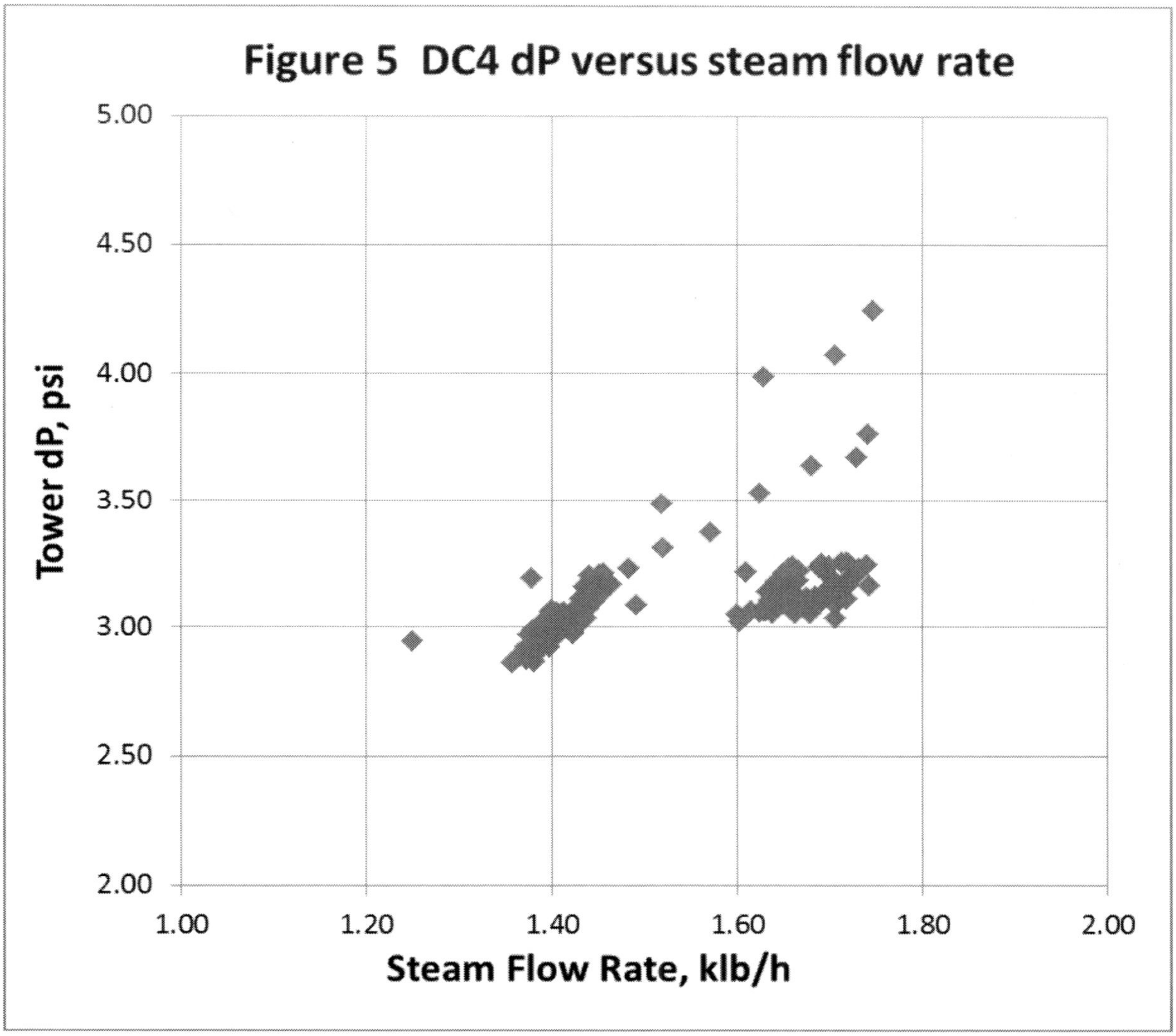

Table 1 summarizes our hydraulic calculations. The calculations do not show proximity of any of the major flood limits. The percent jet flood for all trays using two different methods is below 50%, which is in superb agreement with the gamma scans that showed froth heights less than 50% on all trays in all scans. Downcomer liquid backups are very low, precluding any likelihood of normal downcomer backup flood (assuming absence of fouling or foaming, which is a good assumption here). Downcomer inlet velocities are low (tight is around 0.4 ft/s) on all trays. We concluded that none of these flood mechanisms were likely to limit this tower.

TABLE 1 Hydraulic Calculations for the debutanizer based on the February 17, 2015 test

Trays	Above feed	Below feed
Column diameter, in	32	32
Tray spacing, in	21	18
Active area, total, ft2	4.6	4.5
Downcomer top area, ft2	0.48	1.09
Downcomer bottom area, ft2	0.48	0.48
% open slot area (note 1)	14	15
Weir height, in	1.75	1.75
DC clearance, in	1.38	Note 2
C-factor of active area), ft/s	0.170	0.154
Weir load, gpm/in	2.1	2.0
Tray Rating		
% Jet flood, FRI	46%	47%
% Jet flood, Glitsch Eq. 13[1]	44%	43%
DC backup, in clear liquid	5.2	2.4
DC inlet velocity, ft/s	0.21	0.11
Pressure drop, in liq.	5.5	4.6
Dry pressure drop, in liq.	1.1	1.1
Head loss at DC exit, in liq	0.11	0.33

Notes:

1. Open slot area, expressed as a percent of the active area
2. Truncated downcomers

FOCUS ON DOWNCOMER UNSEALING

The numbers that appear odd are in the bottom row of Table 1, head loss at the downcomer exit. The trays above the feed have conventional downcomers with clearances of 1.38". The liquid heads at the downcomer exits (i.e., through the clearances under the downcomers) were at 0.11 inches of liquid during the February 17th gamma scans, an extremely low value.

Trays below the feed have truncated downcomers, each containing 28 one inch diameter holes. The liquid head at the downcomer exit can be estimated using Chen's[2] method for orifice-type liquid distributors, giving 0.3" liquid head at the downcomer exits, which again is quite low, at the February 17th gamma scan conditions. FRI has an updated version of the orifice coefficient in this correlation which gives a slightly higher 0.33" head loss.

The challenges of handling low liquid loads in tray towers are keeping the little liquid on the tray without drying up, and keeping the downcomers liquid-sealed. When a downcomer loses its liquid seal, gas rises up through it, and interferes with liquid descent. This interference has led to capacity bottlenecks, poor separation, instability, and total inability to start towers up.

On conventional trays, at liquid weir loads above about 2-3 gpm/in of outlet weir length, the outlet weir generates a frothy pool (Fig.6a). As long as the downcomer clearance is lower than the liquid head in the pool, this liquid pool will seal the downcomer and prevent gas rise through it. In contrast, at low liquid weir loads (below about 2-3 gpm/in of outlet weir), and especially at high gas velocities, the gas atomizes the tray liquid, forming a dispersion known as the spray regime or drop regime. In this regime, there is no frothy pool and the liquid resides as atomized drops in the inter-tray space (Fig. 6b). The continuous gas phase reaches the downcomer exit. There is no static seal preventing gas rise up the downcomer. In the spray regime the downcomers are prone to seal loss. Table 1 shows that the weir loads throughout the debutanizer are around 2 gpm/inch of outlet weir. All trays in the debutanizer are therefore operating in the spray regime and are therefore prone to downcomer unsealing flood. .

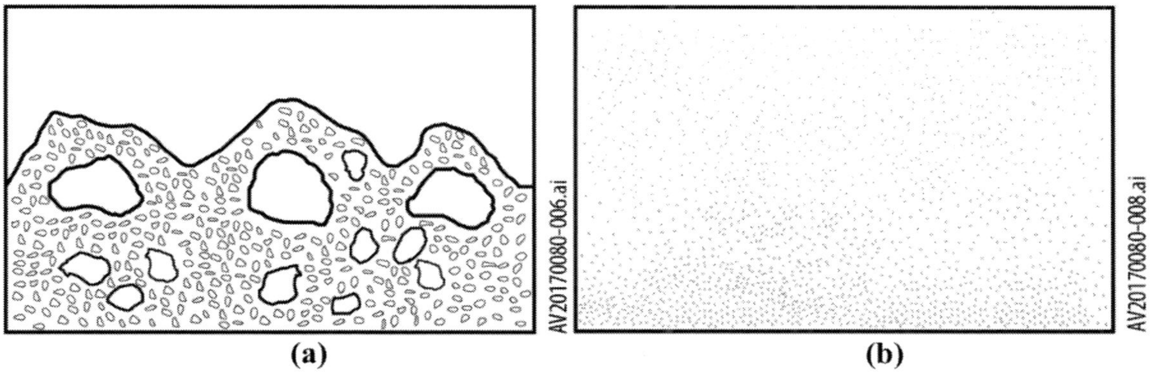

Figure 6. Flow Patterns on trays (a) Froth regime (liquid phase is continuous) (b) Spray regime (gas phase is continuous).

Fig. 7 illustrates how downcomer unsealing can bottleneck a tower in the spray regime. This is a common scenario at startups at low liquid rates. Consider a tray with very little liquid on it. The gas, looking for the easiest path, ascends through both the trays and downcomers (Fig. 7a). Upon liquid introduction, some reaches the downcomer across (Fig. 7b). If the gas velocity through the downcomer is too high, it will not allow the liquid to descend into the downcomer (Fig.7c). Unable to descend, the liquid will accumulate on the tray above. The tray will flood, and liquid will carry over the top of the tower (Fig. 7d).

An alternative scenario develops when the liquid head that accumulates (Figure 7c and d) leads to weeping. The weeping partially seals the downcomer clearance, reducing the vapour velocity through it. The liquid now descends and the flood is relieved. Once the liquid passes, the downcomer unseals again. Instability develops, where a flood step is followed by a dump step. The column would intermittently flood and dumps.

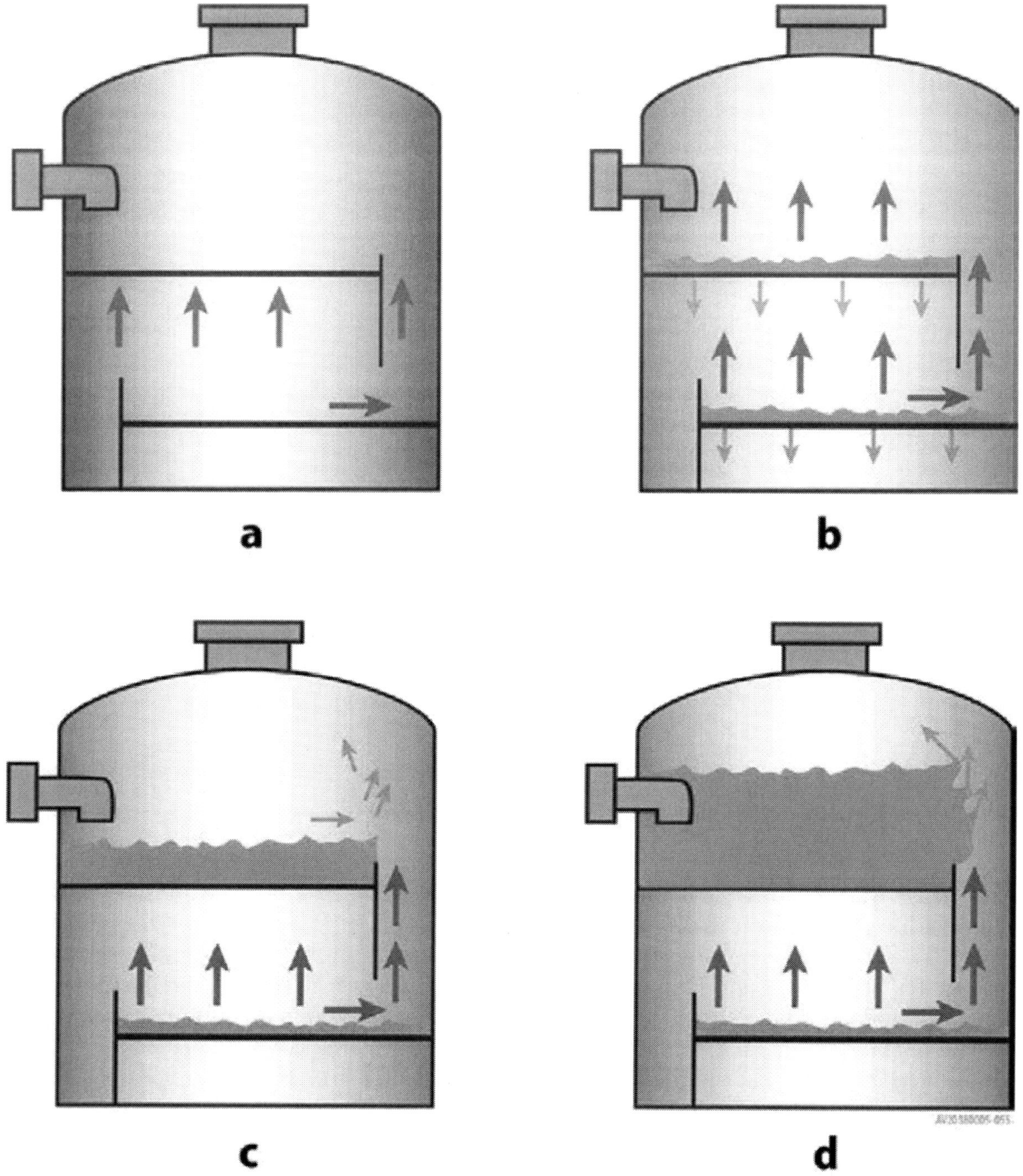

Figure 7. How Downcomer Unsealing bottlenecks a tower in the spray regime (a) Gas distributes between tray and downcomer; (b) Liquid introduced, reaches the downcomer, attempts to descend; (c) When gas velocity up the downcomer is too high, the gas does not permit liquid descent; (d) Unable to descend, liquid accumulates on the upper tray, generating flood. ***(from H.Z. Kister, "Practical Distillation Technology," Course Manual. Copyright © 2013; reprinted with permission.)***

CAN DOWNCOMER UNSEALING FLOOD OCCUR IN THE DEBUTANIZER?

Fluor has a hydraulic model for evaluating the likelihood of downcomer unsealing flood on trays operating in the spray regime. This model was described in the literature[3,4] We applied this model for the debutanizer based on the February 17th, 2015 test loads.

Our calculations show that the vapor fraction rising through the downcomers is about 25% for the trays above the feed and 20% for the trays below the feed. The model predicted that the downcomers of the trays above the feed may experience downcomer unsealing flood when the vapor load exceeds 10,000 lb/h, while these on the trays below the feed may experience this flood at vapor loads exceeding 12,500 lb/h. The velocity through the downcomers was at 166% and 119% of the downcomer unsealing flood on the trays above and below the feed, respectively, based on the February 17th gamma scan tower rates. This means that a downcomer unsealing flood is likely to have occurred both above the feed and below the feed, with the likelihood significantly higher above the feed.

In reality, the calculation is inaccurate, especially for the trays above the feed. Our model assumes that the entire gap between the tray floor and the bottom of the downcomer bar is open for vapor flow. In reality, some of this opening will be covered with liquid. This liquid coverage may be a significant fraction of the relatively small downcomer clearance of 1.38". Our model shows that if there is as little as 0.5" of liquid layer at the tray floor, which is quite reasonable for the spray regime, the percent downcomer seal flood may decline from 166% to 100%, and the flood vapor rate through this section at the downcomer unsealing flood rate will be 15,500 lb/h, as was observed in Figure 4. .

So for the trays above the feed, the very small liquid head loss in the downcomer exit (0.11 inches of liquid) will do little to resist vapor rise through the downcomer. Once the vapor breaks through, these trays are likely to operate above or close to the downcomer unsealing flood point limit.

The situation is different on the trays below the feed. As described above, our model predicts that the vapor will prevent liquid descent through the downcomers. However, there is a sudden expansion of the bottom section downcomers, 5.5 inches above the downcomer outlet holes, where the downcomer area more than doubles. Above the expansion the velocity of vapor in the downcomers becomes 50% of the downcomer seal flood, and the flood will clear, possibly not reaching the tray above. The tray will become a dual flow tray with vapor rising through both trays and downcomers, but no flood propagating upwards. This is somewhat similar to the "blowby" mechanism commonly experienced in truncated downcomers of trays with a multitude of downcomers when the head loss at the downcomer exit spouts is low. This blowby is discussed elsewhere[5]. This phenomenon will reduce tray efficiency, but will not lead to flooding.

In summary, it is likely that downcomer unsealing occurred in most trays of the debutanizer. On the trays above the feed, this unsealing is likely to have led to flood. On the trays below the feed, the unsealing is unlikely to have led to flood, but is likely to have caused blowby and a separation efficiency loss.

BAFFLING BAFFLES

The mechanism above describes all the observations associated with the flood, but is unable to explain the fluctuations. The regular nature of the fluctuations even during unflooded operation points to another mechanism in action in this tower.

Figure 8 shows the operating chart for several key variables during the gamma scan test on February 17th, 2015. During the gamma scan test (10:20-11:20) the liquid level was swinging mostly between 55 and 85% of its range, at a frequency of about every 10 minutes or so, and the bottom flow rate swung between 0 and 3 klb/h from peak to valley with it. The control temperature and the tower dP also swung but at a different frequency than the bottoms level. Following the test, after the pressure drop came down, the swings continued but at a significantly lower amplitude.

Figure 8 Operating Chart for Key variables During the Gamma Scan Test

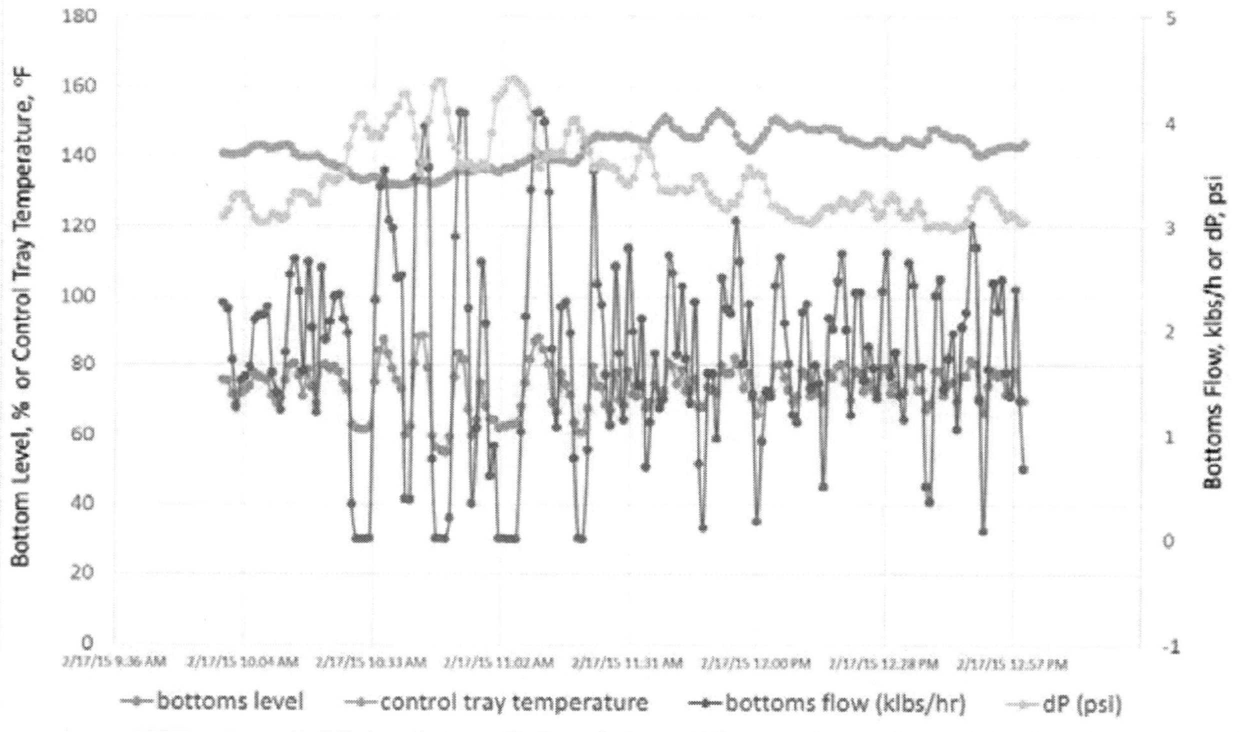

Figure 9A shows the baffle arrangement at the bottom of the debutanizer. The downcomer from the bottom tray is submerged deep within the reboiler draw sump. Liquid from the reboiler return flows to the top of the sump, under an upper sloped baffle, and then overflows a vertical baffle to reach the bottom sump.

There are two issues with this arrangement. First, the liquid in the bottom downcomer is aerated, while that in the reboiler draw sump is not. The non-aerated liquid in the reboiler draw sump has a higher specific gravity than the frothy liquid in the downcomer, which will lift the frothy liquid a fair distance towards the bottom tray. With a typical downcomer liquid aeration of 50%[6], the 5.5 feet submergence will equal the distance from the top of the liquid level to the bottom tray above, and the frothy liquid will be pushed all the way to the bottom tray. In reality, the specific gravity of the frothy liquid in this long downcomer is probably higher than 0.5, so the froth level in the downcomer will be too short to reach the bottom tray as shown in Figure 9A.

This mechanism of the dense liquid in the sump raising the frothy liquid to the tray above is known to have caused bottlenecks and is discussed in item 17, page 110, of *Distillation Operation*[6].

Figure 9B shows the second, and much more important issue. The vapor space above the bottom compartment is not vented. This means that vapor getting in is unable to get out and will back-pressure the liquid in the reboiler draw compartment. When it does this, the amount of liquid overflowing the vertical baffle will decline, and the bottom level will drop. The higher pressure in the vapor space above the bottom compartment will push liquid up the reboiler draw compartment, and its level would rise. This will also raise the backup of frothy liquid in the bottom downcomer towards the bottom tray.

This process will continue until the liquid head above the vertical baffle reaches 2 feet or so, at which time the vapor space above the bottom compartment would reach the bottom of the sloped baffle (Figure 9C). The tower bottom level will reach a minimum, while the frothy liquid height in the bottom downcomer will peak, possibly flooding the bottom tray.

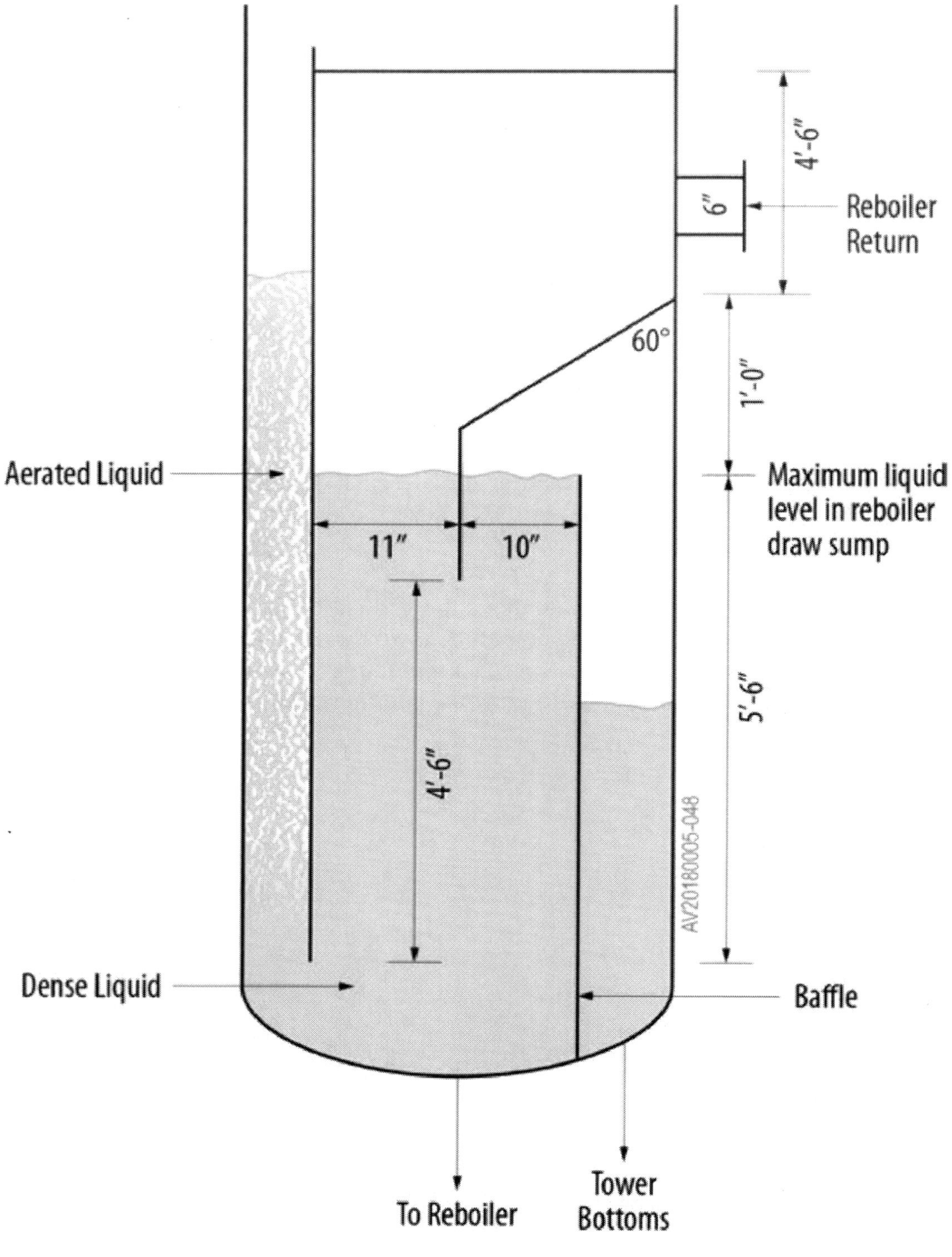

Figure 9. Debutanizer Bottom Sump Cycle (a) Initial Steady State

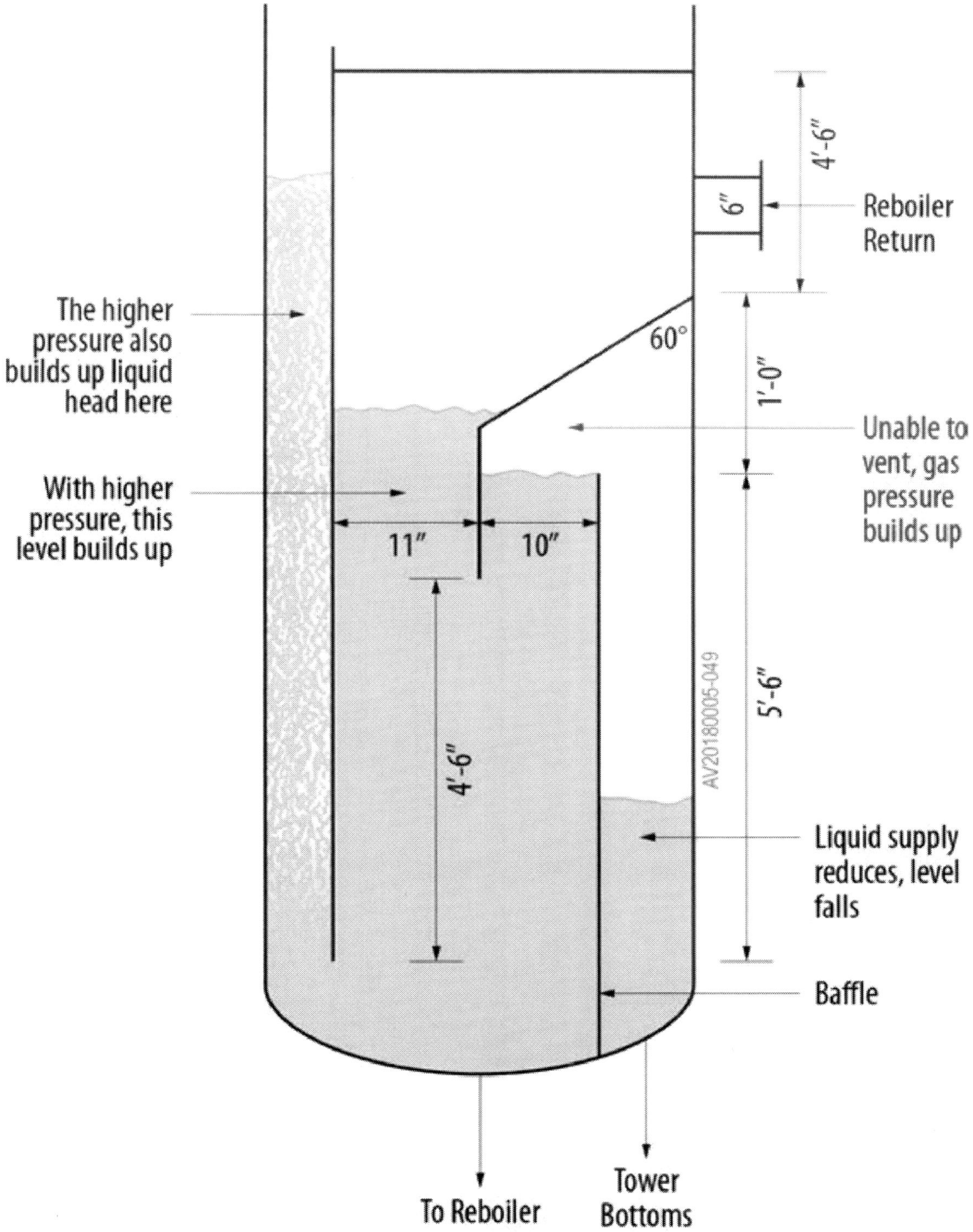

Figure 9. Debutanizer Bottom Sump Cycle
(b) Pressure Buildup

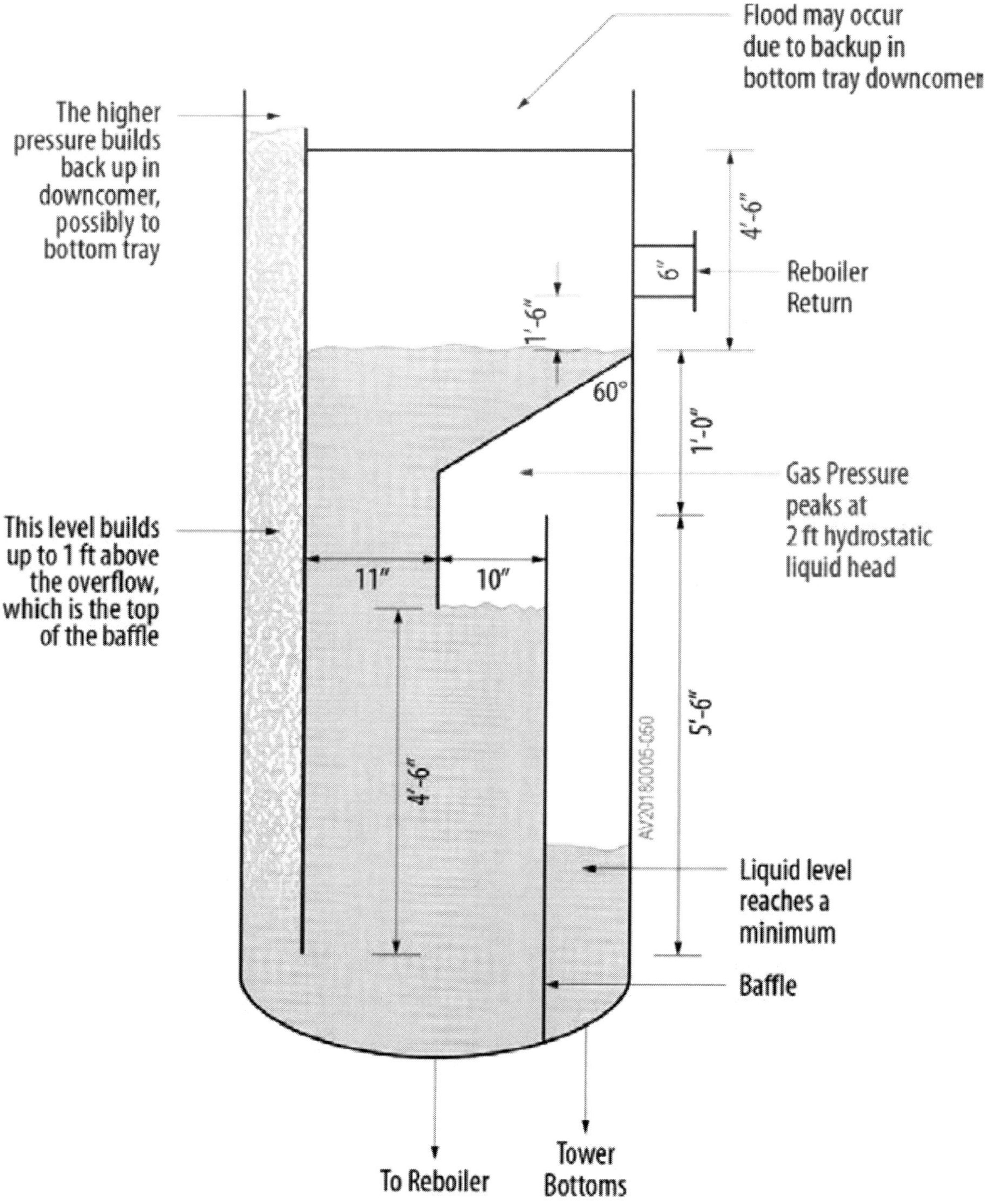

Figure 9. Debutanizer Bottom Sump Cycle (c) Pressure Peaks

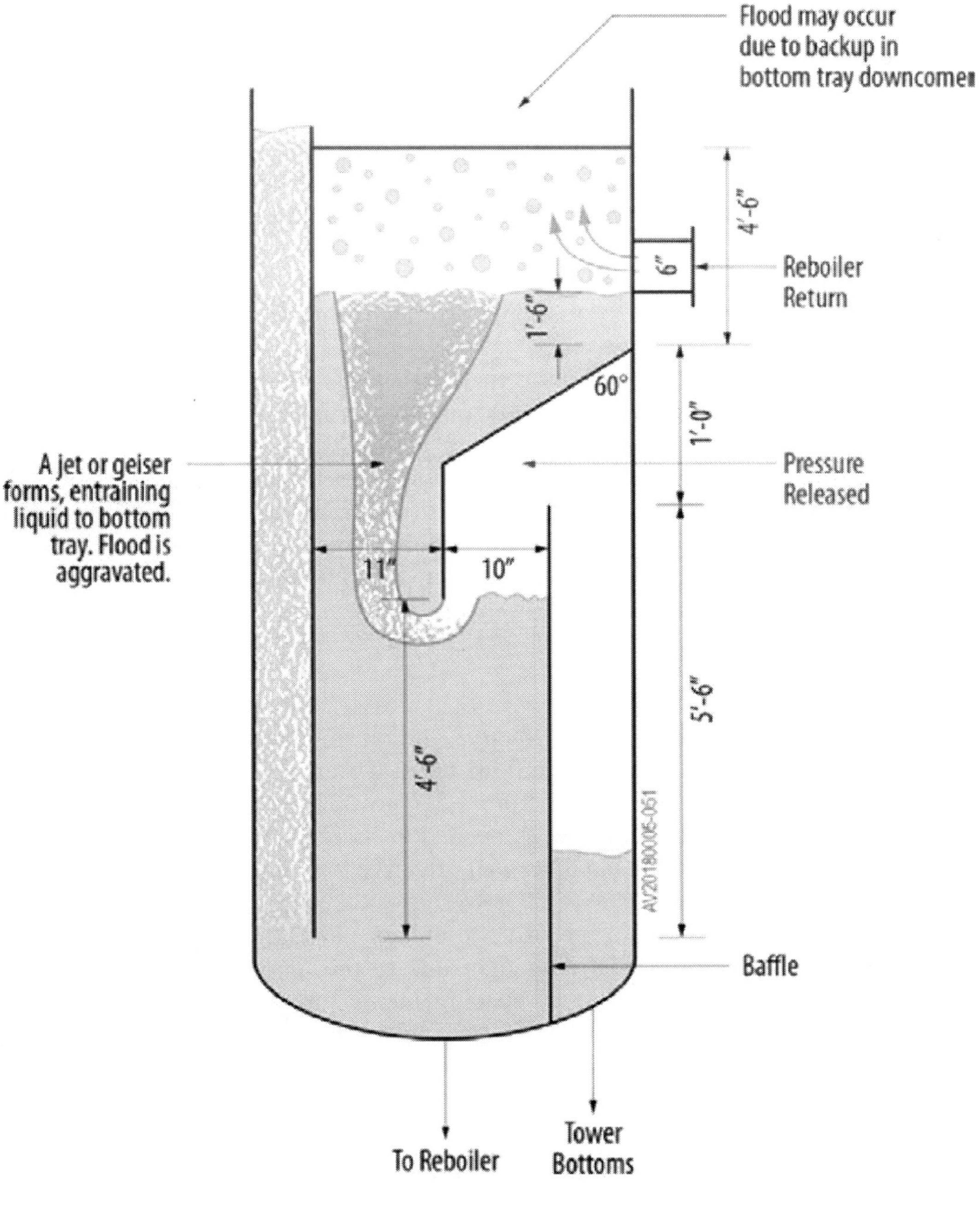

Figure 9. Debutanizer Bottom Sump Cycle
(d) Vapor Breakthrough and Pressure Release

A vapor path now forms and the vapor geysers out through the liquid (Figure 9D). Once the puff is gone, the reboiler draw compartment returns to its initial state in Figure 9A, and the process would repeat. This mechanism is likely to explain the observed fluctuations in the debutanizer.

The gamma scans provide evidence supporting the occurrence of this mechanism. In the February 17th scan, the scanning contractor stated that the liquid level in the reboiler draw compartment was 2 feet below the centerline of the reboiler return nozzle. This is 3 inches below the top of the sloped baffle. Our gamma scan evaluation is that the liquid level was three inches higher than that, right at the top of the sloped baffle. This is the height to which the liquid level in the reboiler compartment would reach according to our postulated mechanism. The liquid level detected by the scans was 9 to 12 inches higher than where it should have been (flush with the top of the vertical baffle) had the baffle system operated like it should.

Also worth noting is that in the February 12th scan the level in the reboiler draw compartment was 6 to 12 inches lower, a variation consistent with our mechanism. Also, the scans show a significant amount of liquid, either entrainment or weep, in the vapor space above the liquid level in the reboiler draw compartment. Entrainment would be consistent with Figure 9D. However, weep is supported both by the flat appearance of the gamma scan peaks and by the presence of this liquid in the vapor space above the reboiler draw compartment also in the previous scans. Weep may be due to some froth backing up onto the bottom tray from the downcomer as illustrated in Figures 9C and 9D.

CURE

To prevent the downcomer unsealing, the corrective action required was to reduce the downcomer exit areas, both of the top and the bottom trays. To achieve this, the downcomer clearances above the feed needed to be reduced from 1.38" to 1" or inlet weirs needed to be added. For the trays below the feed, the spout areas needed to be reduced, by blanking holes at the bottom of the downcomers. This would have been an effective low-cost solution.

For process reasons, there was an economic incentive to enhance separation in the tower. With the trays being hydraulically lightly-loaded (other than due to the unsealing issue) it became attractive to install additional trays at lower tray spacing in the top section. New trays were installed throughout the tower, all of similar design, with no truncated downcomers. The new trays had slightly reduced downcomer clearances of 1.31 inches, as well as 0.75" tall inlet weirs a short distance downstream of the downcomer exits so that the effective downcomer clearances were reduced to the specified 1 inch.

To overcome the venting problem, the solution was to vent the bottom draw compartment using a gooseneck vent (Figure 10), and to shorten the downcomer from the bottom tray, reducing its submergence. This was implemented as specified.

The tower is now operated with better separation than ever at 10% higher loads than previously with no flooding, low pressure drop, and no instability, attesting to the value of good diagnosis for economically solving tower problems.

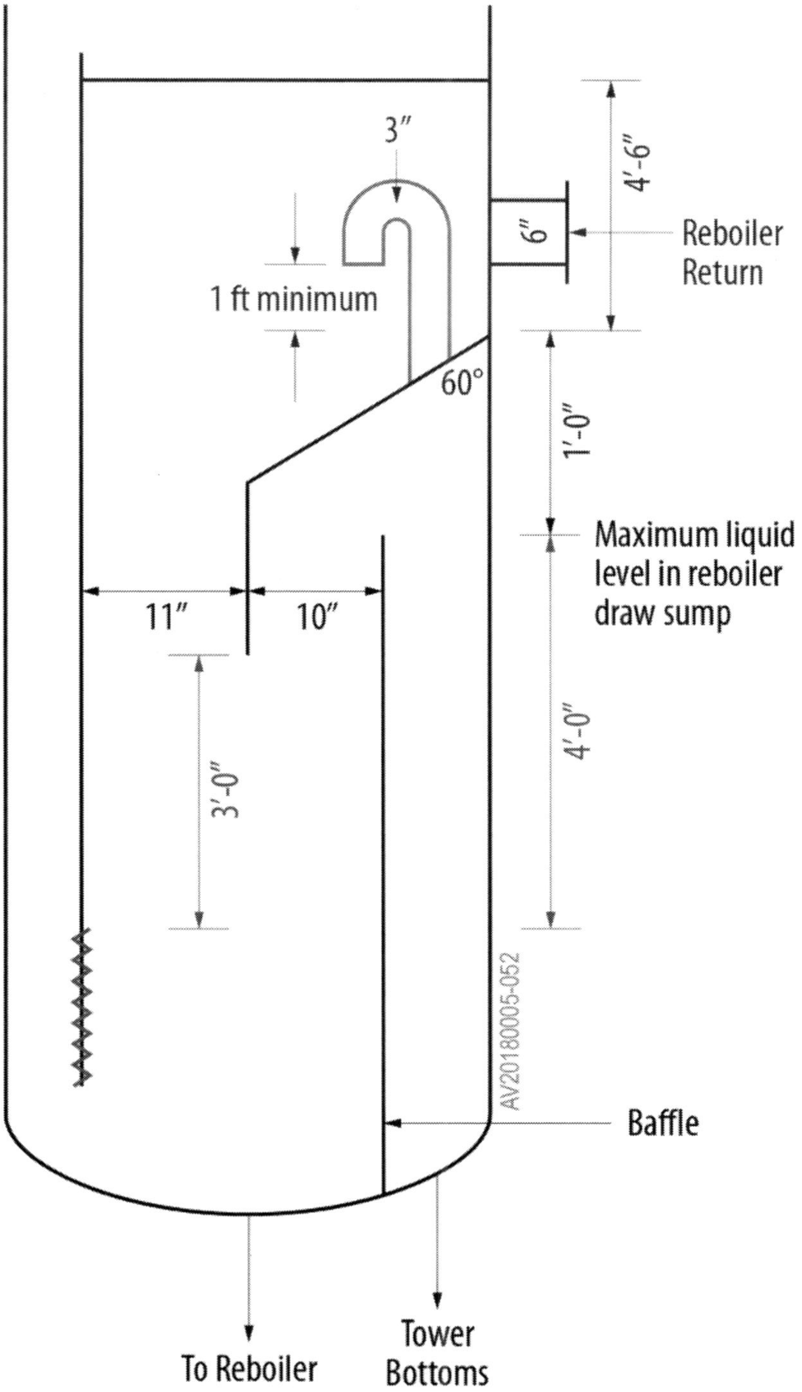

Figure 10. Debutanizer Bottom Sump Baffle Modifications

REFERENCES

1. Glitsch, inc. "Ballast Tray Design Manual", Bulletin 4900, 6th Ed, Dallas, Texas, 1994 (available from KochGlitsch LP, Wichita, Kansas).
2. Chen, G. K., "Packed Tower Internals", Chem. Eng., p.40, March 5, 1984.
3. Kister, H. Z., "When Tower Startup Has problems", Hydrocarbon Processing, 58(2), p.89, 1979
4. Kister, H.Z., and N.M. Mohamed, "Troubleshoot and Solve Gas Treater Downcomer Unsealing Problem: Part 2", Gas Processing, March/April 2015, p.49.
5. Kister, H.Z., "Apply Gamma Scanning to High-Capacity Trays", Chem. Eng. Prog., April 2015, p.45.
6. Kister, H. Z. "Distillation Operation", McGraw-Hill, 1990.

551726 Practical Approach to Distillation Tower Revamp

Whether addressing capacity increases, change in feedstocks and/ or resolving a known constraint, the approach to revamping distillation tower should follow a systematic methodology. A clear problem definition and objectives is imperative for defining the scope of the revamp. Evaluation of the tower performance in its current state and identification of capacity constraint will define the gaps to fill for meeting the revamp objectives. Alternatives can then be evaluated and short listed for framing the scope of the revamp. Those alternatives may revolve around process integration, replacement of tower internals and/ or new tower. Vendor selection would follow, with detail engineering, fabrication of the equipment, installation and commissioning. Those activities will require review of vendor proposals, drawings, shop and field inspections. Technical support would also be provided during commissioning and startup. This paper discusses process engineering activities involved in the different phase of a revamp. Simulation work and gathering of historical data is discussed relative to problem definition and identifying gaps. Data typically supplied to equipment vendors and key parameters reviewed from proposals, drawings are also discussed along with tips for shop inspection and commissioning.

Stable Thermosiphon Reboiler Operation: with or without a Valve in the Inlet?

Edwin van Doorn, Fluor, Amsterdam, The Netherlands

and

Henry Z. Kister, Fluor, Aliso Viejo, CA

Presented at the Kister Distillation Symposium, AIChE Spring Meeting, New Orleans, Louisiana, March 31 - April 4, 2019

Introduction

A reboiler provides the necessary heat duty for the separation in a distillation tower. It boils some of the tower bottom liquid using a heating medium such as steam or hot oil and returns vapor to the tower. The reboiler duty is usually either held constant or manipulated to keep a steady temperature on one of the lower trays. This heat duty sets the vapor load in the tower lower section.

Thermosiphon reboilers are the most widely used reboiler type in distillation systems and are usually considered first[1]. They are relatively compact and economical, require no pumps, and offer relatively high heat transfer rates (small exchanger size) with low residence times in the heated zone.

Design and operation of thermosiphon reboilers were discussed as far back as the 1960's[2,3]. A recent review by the FRI Design Practices Committee[1] focused on the reboiler circuits, the integration of the reboiler with its piping and the tower bottom sump. However, to-date, one key consideration received little attention: incorporating a throttling valve in the reboiler inlet line.

A valve in the inlet allows the operator to throttle the inlet pressure to the thermosiphon reboiler to achieve optimum performance. Yet, around the world there are thermosiphon circuits both with and without a throttling valve at the reboiler inlet. Why is that? What is the need of the valve? How does a flow restriction in the inlet piping affect the performance of the reboiler? And what does this mean for the tower?

The pressure balance in the reboiler circuit and the interaction between fluid flow and heat transfer make designing thermosiphons challenging. The engineer must pay attention to details and understand both static design of the system and the potential dynamic behaviors.

This article intend to provide an overview of stable thermosiphon design and operation with or without a valve in the thermosiphon reboiler inlet piping.

How thermosiphon reboilers work

Thermosiphon reboilers utilize the density difference between the liquid in the tower base (and the reboiler inlet piping) and the mixed-phase fluid in the reboiler and return line to drive reboiler process flow. With circulating reboilers, which are the most common types of thermosiphon reboilers (Figure 1), liquid is drawn from the tower base and unvaporized liquid returns to the tower base. The circulation rate through a thermosiphon reboiler is based on the net liquid driving head, amount of vaporization, and the pressure drop along the circulation loop. Figure 1 shows the pressure balance that sets the circulation rate based on the NLL. A similar balance needs to be performed based on the LLL and HLL

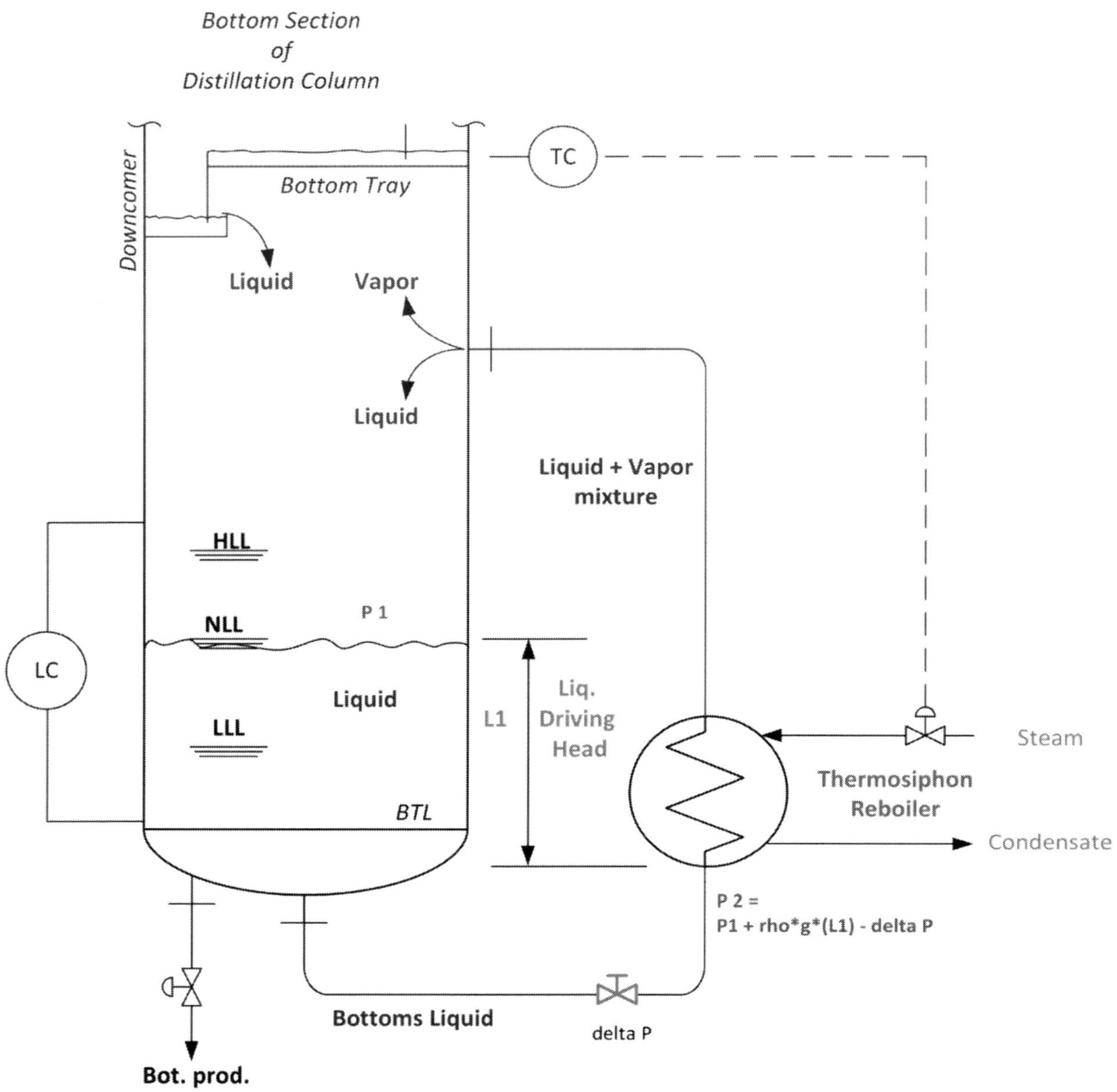

Figure 1. Typical circulating thermosiphon reboiler circuit, shown with a throttling valve in the reboiler inlet line. *This valve may or may not be included, as discussed in the article. Symbols: P is pressure head, Pa; rho is liquid density, kg/m3, g is the gravitational constant, m/s2; delta P is the valve pressure drop, Pa, and L is liquid driving head, m. The numbers 1 and 2 refer to the NLL elevation and the reboiler liquid inlet elevation, respectively. The acronyms LLL, NLL, and HLL refer to the low, normal, and high liquid levels, respectively. BTL is the bottom tangent line, and LC and TC are the level and temperature controls, respectively.*

What impacts the circulation rate?

Regulating the heat duty will impact the circulation rate under the same liquid driving head. Less duty means less vapor generated and heavier density of the liquid/vapor mix; this will alter the whole thermal-hydraulic balance of the system. The system will self-regulate and move to new balance point.

Regulating the liquid driving head will impact the circulation rate under the same heat duty. The effect is not always straightforward and depends on many aspects (e.g. heat flux, pressure in column, bundle geometry, type of reboiler, reboiler/column leveling, natural convection effects, boiling range, mixture effects).

With a condensing heating medium (e.g., steam), the heat duty can be controlled either by manipulating the reboiler steam pressure (Figure 1), which varies the condensing temperature and therefore the reboiler temperature difference, or by holding a liquid level in the heating side, manipulating it by varying the condensate flow rate. In the latter scheme the flooded area does little boiling, which reduces the effective heat transfer area and the net liquid driving head, in turn reducing circulation.

Fig. 2 is a schematic of a tube in a vertical thermosiphon reboiler. The column sump liquid is at its bubble point temperature at the top of the liquid level. At the bottom of the column, the pressure is higher due to the liquid static head and therefore the liquid is subcooled below its boiling point temperature. The first part of the heat transfer area in the reboiler is therefore used for bringing the subcooled liquid back to the boiling point at higher pressure. This is termed “the preheat zone”. This sensible heat transfer has relative low local heat transfer coefficient compared to the boiling heat transfer coefficient.

Once the boiling point is reached, boiling generates a two-phase mix of lower density. The lowest boiling zone is “bubble flow”. The bubbles form by nucleation at the heating surface and are dispersed in the liquid. Upon more heating, more bubbles form, coalesce to larger bubbles, and then to slugs of vapor, termed “slug flow”. Each reboiler tube slugs at a different pace so overall no instability develops. Upon further heating, an annular flow zone forms, where the liquid upflows as an annulus and the vapor flows inside as a core. The vapor shear drags the liquid film up with it. Heat is removed from the wall by convection, with vaporization taking place at the vapor-liquid interface. Further heating boils off the liquid film, the wall dries up, and a mist flow zone sets in. Heat is transferred via a vapor film that has a high thermal resistance, and the tube wall temperature approaches the heating medium temperature. This condition is often termed “burnout”, and may lead to metal overheating, fouling and poor heat transfer, so it needs to be avoided.

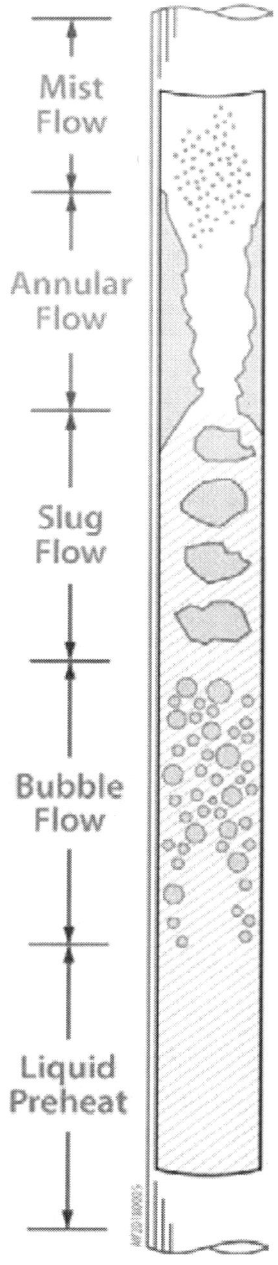

Figure 2. Flow regimes in a tube of a vertical thermosiphon reboiler

In medium/high pressure applications a small pressure change makes very little difference to the boiling point, the subcooling is small, and the sensible heating zone is short. The vapor generation is only slightly influenced by the additional liquid static head. Higher liquid driving head means higher driving force, hence higher circulation and lower vapor percentage.

In low pressure applications the small pressure changes have a much larger influence on the boiling point so the subcooling degree becomes significant and can reduce the duty of the reboiler (and thus the vapor generation). For these services it is important that the reboiler is designed for subcooled liquid at inlet and also that the heat duty process control and the elevation of the thermosiphon are set to prevent unstable interaction between circulation rate and heat duty. In vacuum applications, small changes in absolute pressure make a large difference to the boiling point, so even a small static head may be enough to cause deep subcooling, which, in turn, results in very long heating zone (50% or more) and consumes large portion of heat duty and vapor generation (30% or more). For the same heat duty a higher liquid driving head can result in lower circulation.

Stable circulating flow

It is important to recognize that the heat duty is set to maintain a process requirement and NOT to have desired circulation. Circulation rate only can be achieved by adjusting static head. Proper circulation rate with acceptable vaporization percentage can be achieved by adjusting static heads. The circulation rate is vital to stable column operation.

For circulating thermosiphon reboilers, the best performance is often achieved when circulation flow is maximized -— as long as the loop remains stable. High thermosiphon circulation reduces fouling tendency and often improves heat transfer at medium / higher operational pressures.

Generally, thermosiphon reboilers should be sized for 10-40% vaporization. The exact value depends upon the construction details and the system involved. The lower limit (10%) is to ensure stable thermosiphon flow (i.e. adequate density difference between liquid at inlet and mixture at outlet, adequate/stable pressure drop distribution along the circulation loop). Fluids with large density difference between liquid and vapor (e.g. water) can have a lower limit down to 5%. The upper limit (40%) is to keep a comfortable margin away from mist flow. Typical reboiler design strives for 20-25% vaporization for the controlling design case. As long as all other operating scenarios fall within the 10%-40% range, the thermosiphon loop should be ok.

Stable thermosiphon flow is characterized by the following guidelines:

- Pressure drop in the inlet line should be in the range of 25-40% of the total reboiler circuit pressure drop.
- Liquid velocity in the inlet line should be 0.3-1.5m/s (0.9m/s preferred), but the pressure drop criterion above should dominate.
- The design heat flux (q=Q/A) should be less than 55% of the maximum allowed flux
- Static pressure drop in the outlet line < 35% of the total pressure drop
- Annular flow is required in the outlet piping (in case there is vertical piping)
- Outlet vapor rho-v2 > 60 kg/m-s2 (>100 kg/m-s2 preferred)

- The weight fraction vapor in the range of 0.1 - 0.4

The above rules do not extend to deep vacuum operation or to operation above 0.88 of the pseudocritical temperature of the mixture.

Designed thermosiphon operation versus actual operation in the field

It is impractical / impossible to design any system with 100% perfection. Often there are several margins incorporated during design phase (e.g. margin demanded by end user, overly conservative friction factors for calculating pressure drop in pipes and fittings, etc.). In most of the designs, there are multiple cases involved (various operating cases, turndown, SOR, EOR) and the basis for the final design is always the most conservative case in fouled condition. Sometimes, not all the data are available at the design stage and design engineers apply (conservative) rules of thumb. This all results often in too large reboiler and a too high static head; in the field the real circulation rate through thermosiphon reboilers is often (much) higher than the calculated design value.

All of these reasons result in overdesign (too much heat transfer area) and/or too high static head (available liquid driving head too high). Many end-users will know that for turndown cases, too high static head can lead to percentage vaporization dropping far below 10% and hence unstable thermosiphon action.

Overdesign often results in too large a reboiler and a too high static head at other conditions; in the field the real circulation rate through thermosiphon reboiler is often (much) higher than the calculated design value. Proper startup can be really problematic due to oversizing.

These issues can be further aggravated when preferential baffles are included in the bottom sump or when parallel reboilers are used as demonstrated by a recent case study[4].

Valve in inlet to thermosiphon

The addition of an adjustable resistance in the thermosiphon inlet will force the hydraulic balance to self-adjust to a new, lower circulation flow. This lower circulation flow should translate to higher vaporization fraction, stable annular flow condition in outlet piping and shorter heating zone.

A valve gives the operator the option of dealing with unstable circulation and subsequently unstable column operation. It can serve as a corrective measure, either to accommodate very wide operating conditions, including various turndown scenarios or to compensate for the actual overdesign in the field compared with the calculated sizing during design. Given that the reboiler is always over-surfaced when clean, and more so if the client also specifies a capacity margin of say 10%, unstable thermosiphon flow could always be a realistic problem. Installing a valve is a simple remedy for this.

This valve has been advocated on an "as needed" basis by many literature sources. Lieberman[5] reported a case in which a reboiler heat transfer coefficient almost tripled by throttling the reboiler inlet valve. We too had experiences of improved heat transfer coefficients in existing reboilers upon throttling the valve. Extensive experimental investigations in pilot-scale reboilers[6] also reported large enhancements in heat transfer upon restricting reboiler inlet. Shellene et. al.[7] presented experimental measurements showing how throttling a butterfly valve in the inlet line to a thermosiphon reboiler effectively dampened U-tube oscillations. Butterworth[13] reports rectifying an instability problem in a vertical thermosiphon reboiler by adding an inlet valve. Zygula and Barkat[8] found this valve useful for troubleshooting for excessive circulation rates; performance improvement upon throttling indicates excessive circulation. This reboiler inlet valve has also been advocated on an "as needed" basis by other literature sources[3,9,10].

Not all agree. Quoting McCarthy and Smith[11], "for normal operation, this valve can be a curse rather than a blessing". They argue that even at full opening, the valve can incur excessive pressure drop and limit circulation. We refute this argument on the basis that the good design practice is for the reboiler inlet line to provide 25-40% of the total reboiler circuit pressure drop (see earlier), so the fully-open valve pressure drop should be less than this. Their second argument is that following startup, operators often forget to re-open the valve, imposing high pressure drops. This should really be prevented by proper operation.

Their third argument is a very valid concern. In reactive, fouling, and unstable chemicals systems, this valve can trap reactive materials, polymers, and tars, especially if left partly open. Reduced circulation due to partial closure or partial plugging at the valve can lead to drying out near the top of the tubes. In some reactive chemical systems, like ethylene oxide, such drying up leads to hazardous, even explosive conditions[12]. Polymer growth may hinder valve closure and tarry deposits can catch fire at the shutdown.

Good engineering judgement is required. The valve should be avoided in reactive, fouling, and unstable chemical services as described above. Adding this valve is often expensive. In case of 24" lines or even with 12" lines to the reboiler this valve adds significant costs. The inlet valve also often incurs additional relief requirements for the isolated reboiler loop.

The valve in the inlet piping should be a rotary type manual valve. Often a butterfly valve is chosen because this valve is simple and compact, is quick to open, has a good control, can deal with large flows, has low pressure drop and is low in weight and costs.

However, a butterfly valve cannot guarantee fully isolation of the reboiler for cleaning, as butterfly valves very often leak. If isolation of the reboiler is required while the tower is on line (e.g. in the case of having a second parallel reboiler or a stand by reboiler), installing of a globe valve would be the better choice. It is more expensive, but a gate valve would be required anyway to get proper isolation. In this case the throttling valve becomes the reboiler inlet isolation valve, so the extra cost is only the differential cost between a globe and a gate valve. The relief requirement is unchanged whether using a globe or a gate valve. When it comes to polymer or tars, they may be trapped easier in a globe valve than in a gate valve, so this consideration may argue against the throttling valve.

Heat duty optimization under temperature pinch

The reboiler inlet valve may offer unique benefits when the reboiler experiences a temperature pinch (process temperature approaching the heating medium temperature), especially at near atmospheric or low pressure conditions. The valve reduces the circulation, which may alleviate the pinch. It may be that many of the experiences in which throttling the valve greatly enhanced heat transfer derive from towers operating under temperature pinches.

Figure 3 shows three pinch conditions. In the upper (single component boiling) and middle (wide boiling mixture) diagrams, the temperature pinch is induced by excessive pressure drop in the reboiler circuit. In both, reducing circulation by an inlet valve alleviates the pinch. In the more common wide-boiling mixture scenario (bottom diagram), the pinch is due to the rise in boiling point of the mixture as lights boil off. With this pinch, reducing the circulation by valve throttling raises the boiling point near the tube outlet and aggravates the pinch. In this case there is little incentive to incorporate the valve, and if already there, it should be kept open.

Recommendation for selecting a valve at Thermosiphon Reboiler inlet piping

In most cases stable reboiler operation can be achieved without a reboiler throttling valve. Good engineering practice teaches that the throttling valve is only needed in case flexibility is required to adjust the flow, vary the liquid driving head to stabilize the flow and/or to reduce the subcooled zone in the reboiler to optimize heat transfer, especially in low-pressure systems.

The following guidelines are recommended:

- In reboiler design it is essential to look not only at the design scenarios, but also at alternative operating cases, turndown, SOR, EOR, startups, shutdowns, where the duties are much lower and the heat transfer coefficients much higher.
- If the process operation is quite constant in a narrow range and operation is at medium/higher pressure, a valve is usually not required and can be omitted.
- With a wide-boiling mixture with a temperature pinch due to boiling point rise near the end of the tubes a valve is best omitted.
- Omit the valve where it can contribute to process hazards such as in reactive chemicals, unstable chemicals, and some fouling systems.

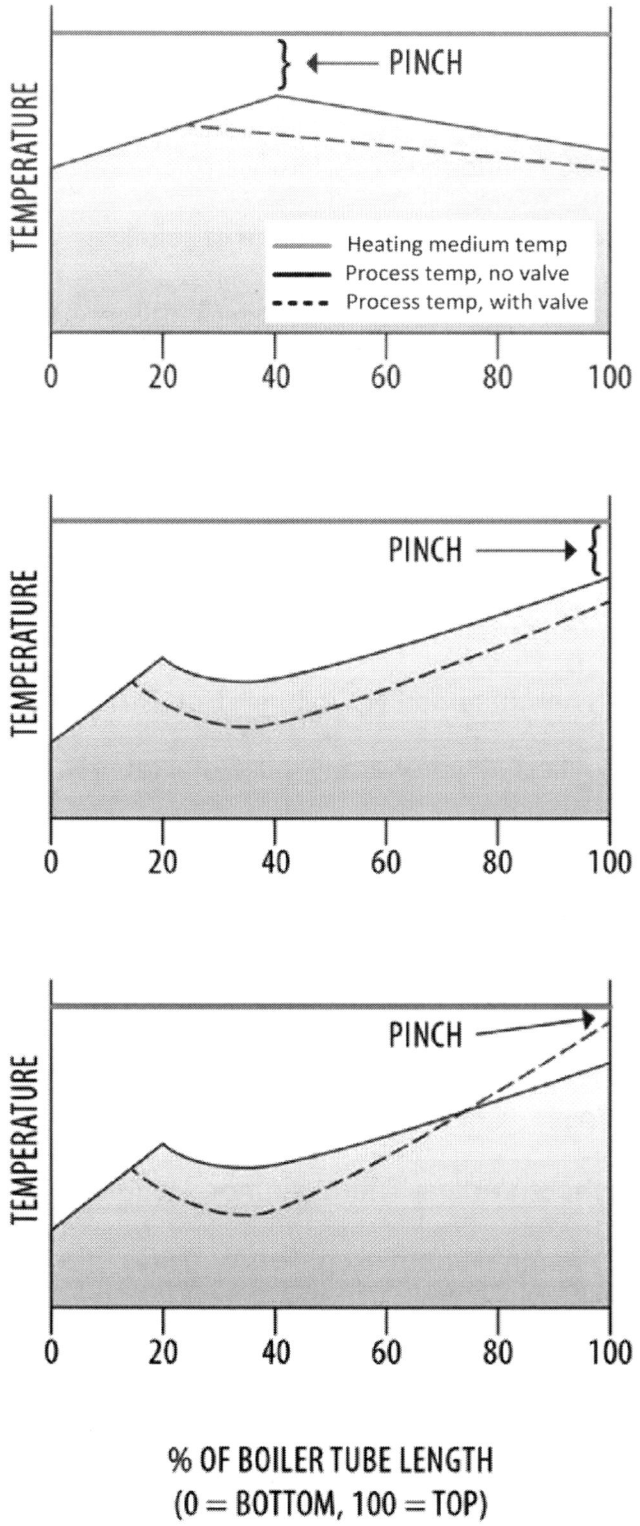

Figure 3. Temperature pinches in a reboiler

- If there is a wide range of possible operating conditions, operating pressure is low / vacuum, exchanger fouling margin is large, available liquid static head is large, long vertical outlet piping is applied, then the valve becomes useful and should be seriously considered.
- An inlet throttling valve is unlikely to harm the thermosiphon loop hydraulics and heat transfer if it is installed and left (locked) fully open.
- The type of valve should be carefully safety-audited and clearly explained to site and operators.
- If adding a reboiler inlet valve, the relief requirements should be carefully evaluated and hazoped.

- Additional considerations apply when multiple reboilers are used or when the bottom sump contains preferential baffles that split it into bottoms and reboiler draw compartments. In these cases the hydraulics can be complicated and must be carefully reviewed.

References

1. The FRI Design Practices Committee, "Reboiler Circuits for Trayed Columns", Chemical Engineering, p.26, January 2011.
2. Fair, J. R. "What You Need to Design Thermosiphon Reboilers", Pet. Ref. 39 (2), p.105, 1960.
3. Jacobs, J. K., "Reboiler Selection Simplified", Hydrocarb. Proc. Pet. Ref. 40(7), p.189, 1961
4. Kister, H. Z.,"Reboiler Circuit Debottleneck with No Hardware Changes", Chem. Eng., August, p. 30, 2017.
5. Lieberman, N. P., "Process Design for Reliable Operation", Gulf Publishing, Houston, TX, 1988.
6. Palen, J. W. ""Shell and Tube Reboilers" in K.J. Bell and associated (Ed.) "Heat Exchanger Design Handbook" vol. 3, rev. 1, Sec 3.6, Hemisphere Publishing, 1985.
7. Shellene, K. R., C. V. Sternling, D. M. Church, and N. H. Snyder, "Experimental Study of a Vertical Thermosiphon Reboiler", Chem. Eng. Symp. Ser. 64(82), p. 102, 1968
8. Zygula, T. M.,and O. Barkat, "Troubleshoot Vertical Thermosiphon Reboilers", Chem. Eng. Progr., July, p.71, 2000.
9. Shah, G. C. "Troubleshoot Reboiler Systems", Chem. Eng. Progr., 75(7), p.53, 1979.
10. Kister, H. Z., "Distillation Operation", McGraw-Hill, NY, 1990.
11. McCarthy, A. J., and B. R. Smith, "Reboiler System Design: The Tricks of the Trade", Chem. Eng. Progr., May, p.34, 1995.
12. Viera, G. A., L. L. Simpson, and B. C. ream "Lessons Learned fro the Ethylene Oxide Explosion at Seadrift, Texas", Chem. Eng. Progr. ,August, p.66, 1993.
13. Butterworth, D. "Heat Exchanger Duty: Going for Gold", Chem. Eng., p.44, December 2004

550219 Kettle Reboiler Circuit Troubleshooting

Kettle Reboiler Circuit Troubleshooting

Mark Cavett, Anand Vennavelli, Chao Wang, Ken McCarley, Tony Cai*

Fractionation Research, Inc.

424 S. Squires St.

Stillwater, OK 74074

ABSTRACT

Reboilers are responsible for many operational problems experienced in distillation and absorption columns in the chemical processing industry. Employed in a commercial size distillation research facility, Fractionation Research, Inc (FRI) kettle reboiler circuits operate under a wide range of process conditions and various physical properties. This makes their operation more challenging. In this paper, operational challenges and problems encountered with the kettle reboiler circuits will be described and discussed. The troubleshooting process and solutions to the operational problems will be presented.

Key Words: Distillation, reboiler, hydraulics, troubleshooting

*Corresponding author. E-mail: cai@fri.org , Phone: 1-405-385-0354, Fax: 1-405-385-0357

A Design Solution to Feed Flow Maldistribution to Multiple Reboilers in Large Diameter Distillation Towers

By

Irene Cullum, Farog Alam
Fluor Ltd., Farnborough, UK
Henry Z. Kister
Fluor, Aliso Viejo, CA, USA

Jan Gertenbach, Michael C Jackson
Tengizchevroil, Farnborough, UK
Christine Long
Chevron, Richmond, CA, USA
Brandon Biega
Chevron, Houston, TX, USA

Christian Bachmann
Thomas Rigert
Sulzer, Winterthur, Switzerland

Presented at the DISTILLATION 2019 VISION 2020: CREATING BETTER DESIGN AND OPERATION FROM PAST LESSONS AND INNOVATIONS, 2019 Spring AIChE Meeting' New Orleans, Louisiana

A Design Solution to Feed Flow Maldistribution to Multiple Reboilers in Large Diameter Distillation Towers

Summary

In refineries, petrochemical, and gas plants it is common to have large diameter distillation towers, in which multi-pass trays are employed and the heat is provided by multiple parallel reboilers. Providing adequate distribution to multiple parallel reboilers has been particularly challenging. Conventional designs have often been plagued with heat transfer limitations, accelerated reboiler fouling, operating instability, transients and swings, and plant bottlenecks.

Recognizing these issues, while designing two large-diameter towers with multiple once-through thermosiphon reboilers, a task force of the owner Tengizchevroil (a consortium of Chevron, ExxonMobil Lukoil and the Kazak partner KazMunayGas(KMG)) and the contractor Fluor (part of the KPJV joint venture) engineers was assembled to develop a design for a collector tray, gathering the liquid from the bottom 4-pass tray, that will provide equal liquid distribution to the reboilers over the full operating range of 40 to 110% of design flows, permit full production rate even when one of the reboilers is removed for cleaning with equal distribution to remaining reboilers, completely decouple liquid distribution from hydraulic disturbances in the reboiler circuits to positively prevent swings and transients, handle reasonable tilt, handle reasonable maldistribution on the 4-pass tray above, and provide good gas distribution to the tray above.

The MOTR (Multiple Once-Through Reboilers) Collector-distributor design developed consists of an internal liquid reservoir maintained by an annular shroud with notches cut in the shroud. Liquid from the inner reservoir overflows the rectangular notches to an annular sump ring. The annular sump ring is divided into reboiler feed compartments by segmenting radial baffles. The number of annular sump ring compartments is a function of the number of reboiler draw nozzles required. Each compartment feeds one reboiler draw nozzle.

The liquid level in the inner reservoir is set by the liquid flow rate and heights of the notches. The annular sump ring floor elevation is well-below the floor of the inner reservoir, so the liquid overflowing the notches falls through the notches onto the annular sump ring. In operation there is no liquid continuity via the notches between the inner reservoir and annular sump ring. The annular sump ring bottom deck is level with the bottom of the reboiler draw nozzles to prevent stagnant liquid hold up on the tray.

Just like any design, the details will make or break it. With the challenging design at hand, a multitude of considerations exist, and a failure to adequately address any of these details can make the difference between success and failure.

To the best of our knowledge, our innovative improved design has not previously been applied. Therefore our team decided to audit the design using CFD and water-test it. The water test was performed at the Sulzer shop in Tulsa, OK, on a full scale prototype of the collector tray design for each of the two columns. Both the CFD audit and the water tests demonstrated that our design met all the objectives described above. For once, a design is available that permits even distribution of liquid to any combination of multiple parallel once-through thermosiphon reboiler circuits even when one is out of service for maintenance.

Our paper provides detailed description of the new MOTR splitter as well as the lengthy list of considerations that need to be addressed to achieve a successful design. Our paper also describes the CFD audit and the water test and its results. It is believed that this development will help engineers struggling with similar issues to achieve good liquid distribution and trouble-free operation of multiple reboiler circuits.

Circulating versus Once-Through Thermosiphon Reboilers

In refineries, petrochemical, and gas plants it is common to have large diameter distillation towers, in which multi-pass trays are employed and the heat is provided by multiple parallel reboilers.

Thermosiphon reboilers are the most widely used reboiler type in distillation systems and are usually considered first[1]. They are relatively compact and economical, require no pumps, and offer relatively high heat transfer rates (small exchanger size) with low residence times in the heated zone. Thermosiphon reboilers (Figure 1) usually vaporize 10-30% of the entering liquid. The differential between the liquid density at the inlet line and reboiler draw sump and that of the vapor-liquid mix in the reboiler and its outlet lines drives the circulation of the mixture through the reboiler. The reboiler outlet vapor-liquid mix returns to the tower.

Thermosiphon reboilers can be the circulating or the once-through types. In a circulating thermosiphon reboiler (Figure 1a) liquid is drawn from the tower bottom sump, and the reboiler outlet vapor-liquid mix is returned to the same reboiler draw sump, so that any unvaporized liquid returns to the reboiler draw sump. In contrast, liquid to a once-through thermosiphon comes from a total draw in the tower, which is either a downcomer trapout, or a collector tray (the collector tray arrangement is shown on Figure 1b) or a baffled compartment in the bottom. The total draw captures all of the liquid leaving the bottom tray and feeds it directly to the reboiler. The reboiler return is directed to the tower bottom and all its liquid drawn as bottoms product. None of the reboiler return liquid is recycled back to the reboiler, hence the name once-through. Trapout arrangements are generally limited to simple, single draw configurations; multi-draw configurations such as from two or four pass trays are best handled with collector trays[1]. The benefit of trapout/collector tray vs a baffled sump is gravity assists movement of the fluid through the reboiler, instead of hindering it. This is of particular importance during cold start conditions when a baffled sump would need to be charged with a “live feed” rather than a stabilized product.

Once-through reboilers have limitations but also offer unique advantages. With a thermosiphon system, reboil vapor is limited to about 40% of the bottoms product rate, due to the normal limitation of 30% maximum vaporization (by weight) in thermosiphon exchangers [1,2]. This makes once-through thermosiphon systems appropriate only for low-boilup systems such as strippers. When the desired vapor boilup rate exceeds about 40% of the bottoms product rate, a recirculating reboiler needs to be used instead.

The advantages offered by once-through thermosiphon reboilers over their circulating counterparts include their achieving one full theoretical stage of separation, a lower boiling temperature where wide boiling range liquids are involved, and often a better temperature difference for the reboiler (as it does not recirculate the hot reboiler return liquid). However most important for large columns is that the high elevation of the collector draw generally provides good driving force for thermosiphon flow, thus minimizing the skirt height. In large towers, shorter skirts translate into handsome savings in capital, making a collector tray arrangement particularly advantageous, and the preferred type whenever they can be used.

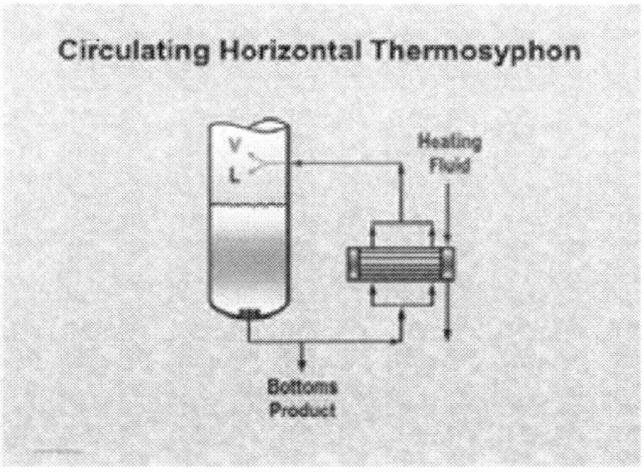

(a)

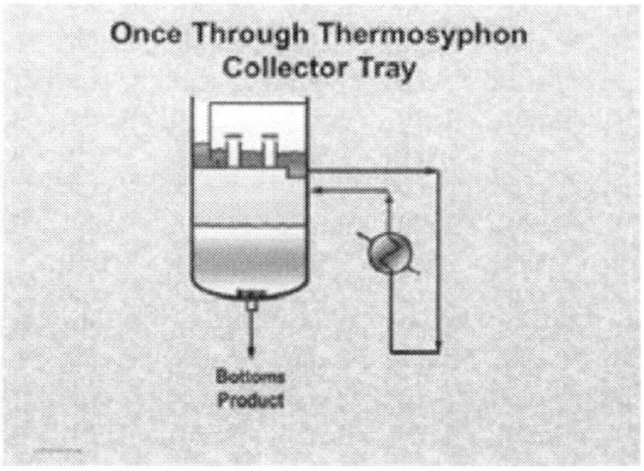

(b)

Figure 1. Horizontal Thermosiphon Reboiler Arrangements. a. Circulating b. Once-through

For these reasons, our article will focus on collector trays for once-through thermosiphon reboilers that provide boiling for large, stripping-type columns.

Once-Through Reboiler Circuit Requirements

To avoid operating problems and ensure optimum performance in systems with multiple reboilers, liquid from the collector must be adequately distributed to all operating reboiler circuits under the entire expected range of operating conditions. This requirement is exacerbated by the use of multi-pass trays and potential maldistribution arising from the distillation section trays above the reboiler draw tray. A recent case study described[3] a heat transfer limitation, operating instability, and a plant bottleneck, in a once-through 4-reboiler system arranged in two parallel circuits, all resulting from maldistribution in the parallel reboiler circuits. The case study further taught that it is important to ensure adequate distribution to the reboiler circuits not only under normal operation, but also at low-rate operation. We are familiar with other cases in which maldistribution between four parallel once-through thermosiphon reboilers introduced instability and temperature swings between the reboilers which translated into troublesome column upsets.

In addition, in the upstream oil and gas industry, due to the nature of the fluids, reboiler fouling is of concern and operators require the ability to isolate and clean individual reboilers while maintaining plant production. This adds a further requirement to incorporate flexibility to evenly distribute the flow to all 'in service' reboilers when one or more of the reboilers are not in service.

Frequently the design employs symmetrical reboiler layouts to try to achieve equal feed distribution of liquid from the collector to the multiple reboilers. Liquid from the multi-pass trays above is fed through downcomers to the reboiler draw collector, which is a chimney tray which may be slightly elevated across the central line, right under the centre downcomer and parallel to it, sloping outwards to the reboiler liquid draw nozzles located on opposite sides of the central downcomer. This is a crude approach to providing good distribution and does not address:

- Flow distribution fluctuations interacting with reboiler circuit pressure drop to generate hydraulic and/or temperature swings and transients back and forth through the reboiler system. Such swings and transients may be unpredictable, often happen unexpectedly, and may persist for lengthy periods.
- Maldistribution due to level tolerances (tilt) on a large diameter draw trays.
- Maldistribution at turndown, when the liquid-split hydraulic may be unfavourable
- Flow maldistribution in the downcomers from the trays above. This risk increases with the number of tray passes.
- The need to redistribute the flow when one reboiler has to be taken out of service for maintenance and cleaning while other reboilers are in operation.

This paper addresses the design principles employed to achieve equal distribution to multiple reboilers in two large diameter towers, one with four parallel horizontal thermosiphon reboilers, and the other with three parallel horizontal reboilers. This design is robust, positively prevents the interaction that generates swings and transients between reboiler circuits, incorporates the flexibility to distribute the flow equally to the active reboilers at high rates, low rates, and when one or more of the reboilers is isolated for maintenance.

Process Description

This process includes a large diameter Crude Stabilizer which essentially acts as a stripper followed by a similarly large refluxed Gasoline Splitter.

CRUDE STABILIZER The Crude Stabilizer is a 6.5 m diameter stripping tower whose purpose is to drive off the light ends to meet the crude bottoms product vapor pressure specification. The tower consists of 4-pass trays. The feed is introduced on the top tray. The stabilizer has 3 steam reboilers, and two heat recovery reboilers. There are four reboiler draw nozzles, located at 90 degrees apart. Each steam reboiler is fed by one nozzle, while the fourth nozzle feeds the two heat recovery reboilers. All are horizontal thermosyphon reboilers fed from a single draw tray.

When all four reboiler circuits are in operation, with a perfectly level reboiler tray and no flow maldistribution in the downcomers from the trays above, each reboiler circuit should receive 25% of the flow.

Where one of the steam reboilers is out of service and all other factors remain the same, it is expected that the crude, rather than distributing evenly to the remaining three reboiler circuits, is split 25% to the heat recovery reboiler and 37.5% to each of the adjacent steam reboilers.

In this design a minimum exchanger overdesign of 50% is required for each of the reboilers to accommodate full production when one reboiler is out of service. With a balancing draw tray, this required overdesign reduces to approximately 33%. A margin is also required to accommodate maldistribution resulting from out-of-levelness of the accumulator tray or maldistribution from the 4-pass active trays. Thus, we sought to develop a robust, flexible design that would avoid swings and transients and achieve even distribution to all 'in service' reboilers, whatever the operating mode, and be robust enough to handle reasonable mal-distribution from the distillation trays or due to tilt on the reboiler draw tray tilt.

GASOLINE SPLITTER The Gasoline Splitter is a 6 m diameter distillation tower whose purpose is to segregate the gasoline fraction which contains the C_1 and C_2 mercaptans from the bulk crude stream. The tower consists of 4-pass trays below the feed. The tower has three steam reboilers, each capable of providing 50% of the tower heat duty. There are 3 reboiler draw nozzles, 120 degrees apart, each feeding a single reboiler.

Flow Distribution to Multiple Reboilers: Conventional Designs

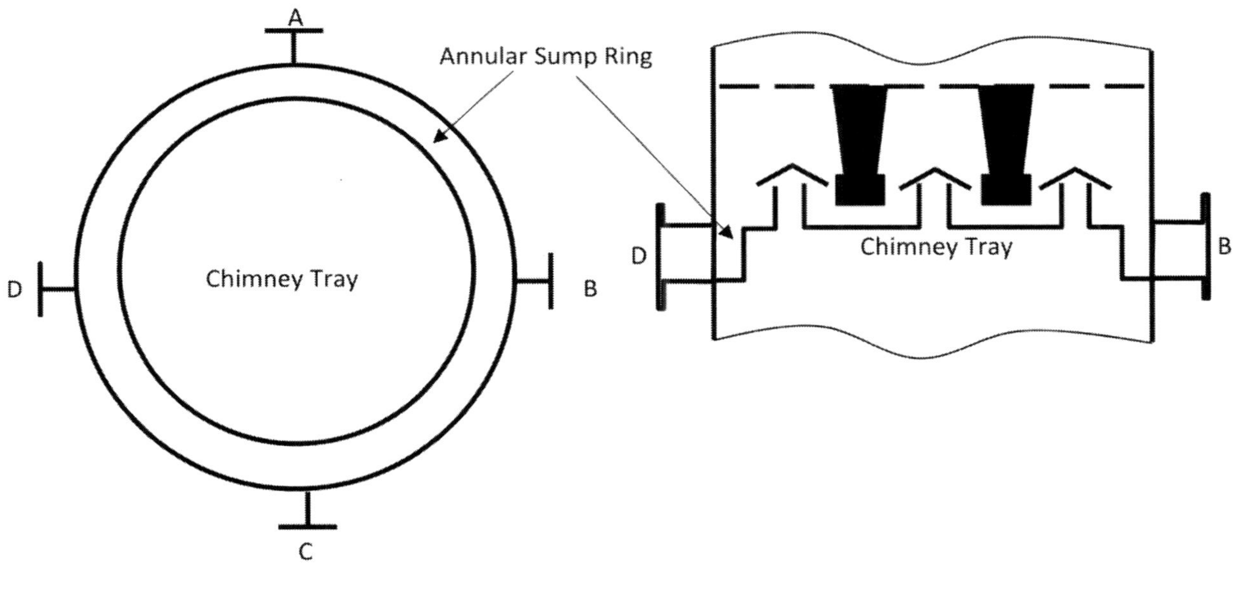

Figure 2: Crude Stabilizer Reboiler Draws

Figure 2 shows a typical chimney tray distributing liquid to four reboiler circuits, with

- Liquid from the off-center downcomers of a 4-pass tray fed to the chimney tray.
- The liquid from the tray flows from the downcomers to the tray annular sump which feeds the four draw nozzles (A to D) that route the liquid to four separate reboiler circuits.

In this case, minimal liquid level is held on the tray or its sump, and the liquid passes directly to the reboiler feed piping. If the chimney tray is perfectly level, and there is no pass maldistribution on the trays above, then all four nozzles will receive equal amounts of flow (25% each). On a large tower, it is very difficult to achieve perfect levelness of the chimney tray, distribution of liquid on 4-pass trays is often far from perfect, and it can be expected that one of the draw nozzles will receive an increased feed.

The flow maldistribution problem is aggravated when one reboiler circuits is taken out of service. For example when the reboiler associated with nozzle A is taken out for maintenance. In this case, the feed piping to Reboiler A will flood and Nozzle A will no longer draw liquid from the chimney tray. The liquid flow direction on the chimney tray will be uneven and flow across the tray is no longer symmetrical to the remaining 3 nozzles, introducing an uneven split to the remaining reboilers.

A flow or temperature disturbance in one of the reboiler circuits can alter the pressure drop in that circuit, which may increase or decrease suction on the liquid entering that circuit. This may alter the quantity of liquid entering all circuits, and this disturbance may propagate back and forth causing swings and transients in the reboiler circuits.

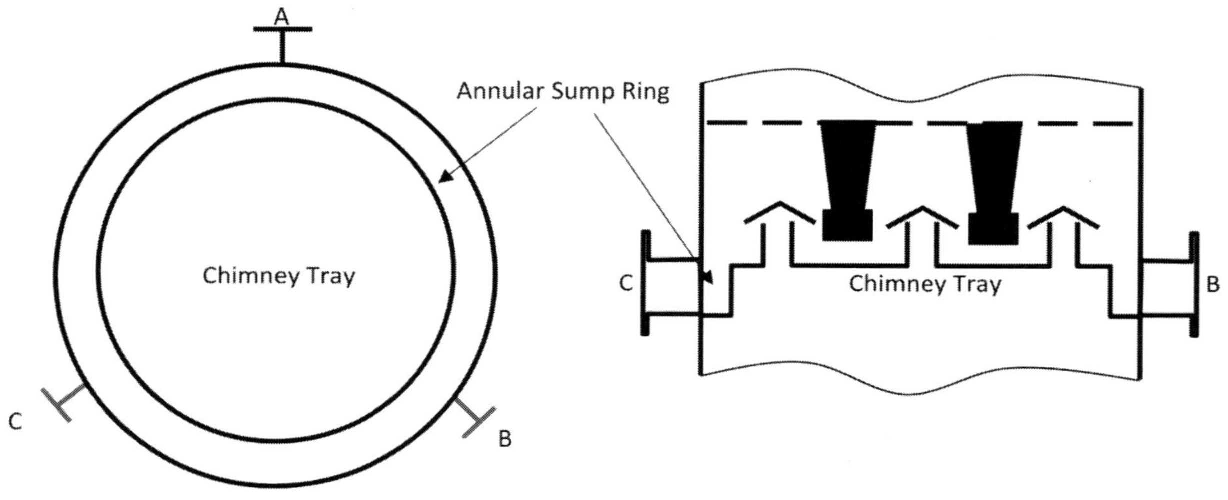

Figure 3: Gasoline Splitter Reboiler draws

Figure 3 shows an equivalent conventional chimney tray distributing liquid to three reboiler circuits from three draw nozzles (A, B, C) at the chimney tray annulus. As in Figure 2, minimal liquid level is held on the tray and the liquid passes directly to the reboiler feed piping. If the chimney tray is perfectly level, and there is no pass maldistribution in the trays above, then all three nozzles will receive an equal amount of flow (33.3% each). As in Figure 2, on a large tower, perfect levelness of the chimney tray and perfect distribution in 4-pass trays are elusive, resulting in maldistribution to the reboiler circuits. The problem will be aggravated when additional unevenness is generated by one reboiler circuit being out of service and when the remaining reboilers are stretched to supply the design heat duty. Likewise, the potential for swings and transients are also issues in the Figure 3 design.

Principle of Improved Liquid Splitting to Multiple Once-Through Thermosiphon Reboilers

In satisfying the requirement to ensure equal distribution to all operating reboilers and thus maintain the production target under the following cases, a range of designs were explored before selecting the preferred design. The following requirements shaped the design:

- Equal distribution over the full operating range of the tower 40 to 110% of design flow
- Full design production rate of the tower to be maintained even when one of the reboilers is removed for cleaning, avoiding lost production when a reboiler is cleaned.
- Complete decoupling of liquid distribution from disturbances to the reboilers and reboiler piping to positively prevent swings and transients.
- Design should accommodate a small specified tilt on the tray and maintain equal distribution to all 'in service' reboilers'.
- Design should accommodate a limited specified maldistribution through the downcomers from the distillation section.
- Design should accommodate equal distribution between all active reboiler draws with at least one reboiler isolated.
- Design should ensure adequate gas riser area (which does not impede liquid flow across the tray) and good gas distribution to the multi-pass trays above.
- Downcomers from the four pass trays above must be sealed.

The MOTR splitter design concept developed here (Figure 4) is to provide an internal liquid reservoir maintained by an annular shroud with notches cut in the shroud. Liquid from the inner reservoir overflows the rectangular notches to the annular sump ring. The flow is essentially metered/distributed by the notches. The annular sump ring is divided into reboiler feed compartments (sumps) by segmenting radial baffles. The number of annular sump ring compartments is a function of the number of reboiler draw nozzles required. Each compartment feeds one reboiler draw nozzle. There is no communication between compartments.

Liquid from the fractionation 4-pass tray above is fed via off-center downcomers (Figures 2 and 3) to the inner reservoir of the reboiler draw tray (Figure 4). The inner reservoir is surrounded by an annular shroud, which has notches cut in it. The notches are closed at the top for mechanical strength, so they look like rectangular slots, but are designed so the liquid level of the internal liquid reservoir does not approach the top of the slots, so they always act as notches. The liquid level in the inner reservoir is set by the liquid flow rate and heights of the notches. The annular sump ring floor elevation is well-below the floor of the inner reservoir, so the liquid overflowing the notches falls through the notches to the annular sump ring whose level is below the inner deck. In normal operation there is no liquid continuity via the notches between the inner reservoir and annular sump ring.

Only when a reboiler is removed from service, the annular sump ring compartment serving that reboiler will flood such that the liquid level in the isolated reboiler compartment equals the level of the liquid in the inner reservoir, thus the liquid in the inner reservoir would now be re-distributed between the remaining active outer compartments and equally feed the remaining reboilers.

The discontinuity between the liquid levels in the annular sump ring and the inner reservoir breaks the coupling that is the root cause of swings and transients. With our improved design, the inner reservoir shroud balances flow to each reboiler circuit, while the liquid level in the inner reservoir sets the ultimate flowrate to each reboiler circuits because the notches control the flowrate from the inner reservoir to the outer annulus compartments. The annulus and reservoir liquid levels are independent of each other, so any feedback of disturbances from the reboiler circuits is not transmitted to the reservoir and therefore does not unsettle the liquid distribution. This prevents the swings and transients experienced with many conventional designs.

In the placement of the chimneys on the draw tray it is imperative that there is sufficient free liquid space to provide numerous channels and sufficient space to avoid obstruction of the liquid flow to the annular shroud notches, with minimal hydraulic gradient. In addition, the placement of the chimneys needed to provide even vapour distribution to the 4-pass tray above taking into account obstruction by the downcomers.

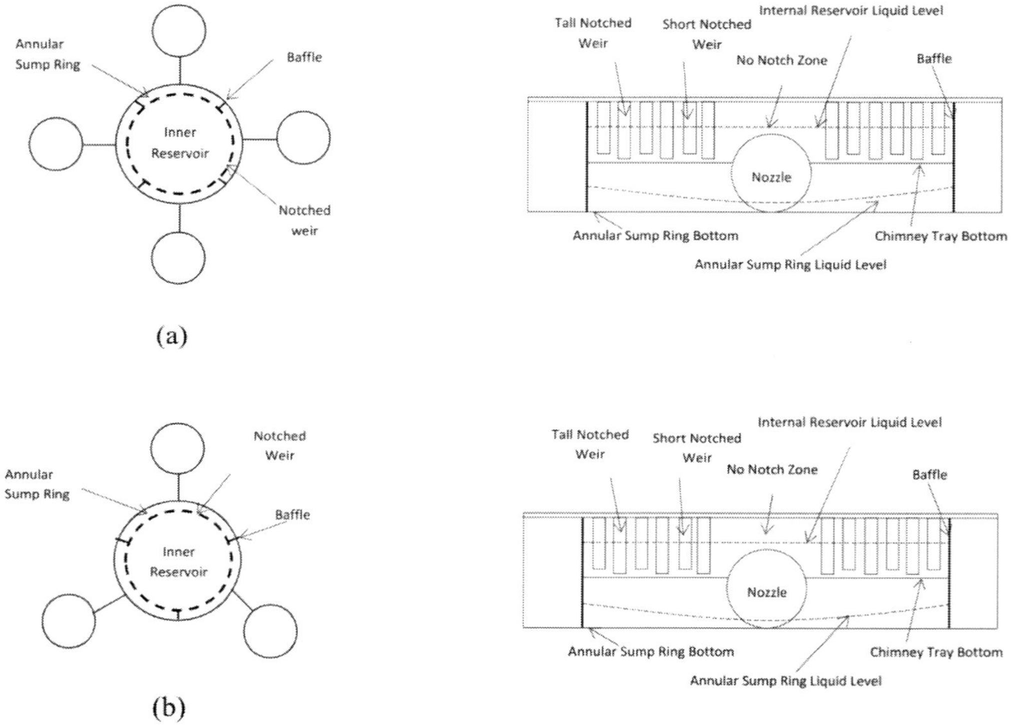

Figure 4: Custom-Made Distributing Chimney Tray Design, (a) Stabilizer (b) Gasoline Splitter

In this design the equal flow distribution is set at the origin and does not rely on the downstream geometry. The design for the crude stabilizer which has 4 reboiler draw nozzles is shown in Figure 5 below.

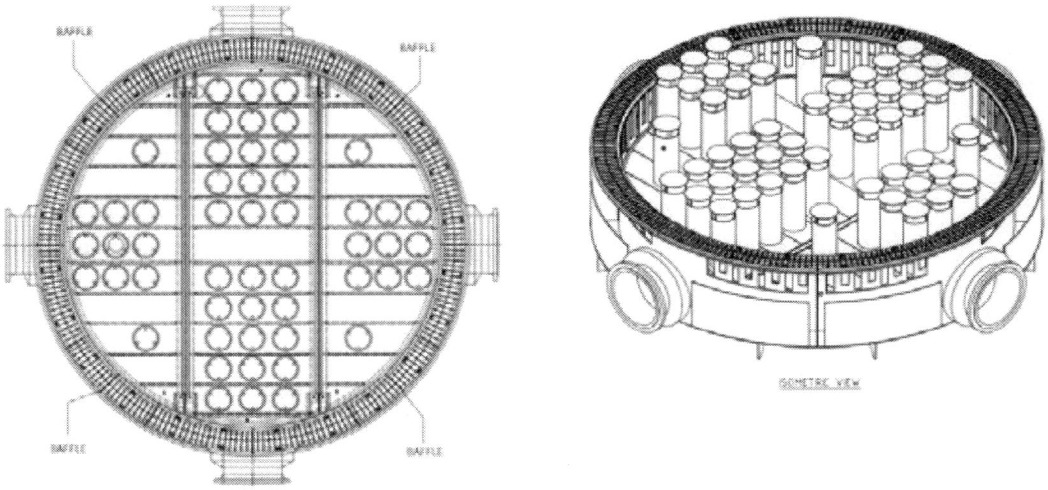

Figure 5: Sketches of the Custom-Made Distributing Chimney Tray Design for the Crude Stabilizer

The Devil is in the Details

Just like any design, the details will make or break it. Overlooking what may seem a minor detail has bottlenecked many a tower[5]. The Fluor-Chevron task force paid special attention to the following key considerations while developing the MOTR splitter design:

ANNULAR SUMP RING

- For the design of these two towers the annular sump ring area is equally divided using radial partition baffles. Thus any liquid flowing into a specific segment of the annular sump ring is routed to the associated reboiler feed piping only. In future designs unequal partitioning of the outer annular ring (combined with uneven notch widths or number of notches feeding each outer compartment) can be used to achieve desired unequal flows to multiple reboilers. It could also be considered in the future to adjust positioning of the reboiler draw nozzles to better suit plant layout restrictions.
- Elevation of the annular sump ring bottom is adjusted to align with the bottom of reboiler draw nozzles.
- Annular sump ring width was sized to ensure adequate vapour disengagement, to avoid restriction to liquid flow towards the draw nozzle, and to minimize hydraulic gradients. For this design, a maximum horizontal liquid velocity 0.4 ft/s was allowed in the annular sump ring.
- The depth of the annular sump ring below the notches of the annular shroud was sized adequate to prevent frothing or foaming from restricting the passage of liquid through the notches.
- Draw nozzles were sized to ensure self-venting including when one nozzle/reboiler is out of service.

INNER RESERVOIR

- The inner reservoir was designed to provide a high enough liquid level in the inner reservoir over the full operating range of the tower. This liquid level ensures good distribution of liquid to the annular sump ring reboiler feed compartments, overcoming the impact of off-level tray levelness and downcomer maldistribution.
- The inner reservoir shroud has a notch design with two different notch heights to accommodate the full operating range (capacity) of the tower including turndown.
- Elevation of the inner reservoir bottom is adjusted such that it is an adequate height above the maximum liquid level in the annular sump ring to avoid any communication from annular sump ring to inner reservoir.
- The notches and volume / height of the annular shroud were sized to accommodate the full flow range (rated to turndown) for the tower without approaching the top of the notch. This way the slots would always act like rectangular notches.
- A drain hole is provided for each compartment to ensure complete drainage of the inner reservoir prior to maintenance.

NOTCHES

- To ensure that sufficient time is provided for vapour disengagement of liquid falling into annular sump ring prior to entering the reboiler feed nozzle, no notches are provided in the inner annulus ring in front of the reboiler draw nozzle.
- The calculations for maximum liquid height in the inner reservoir are based on clear liquid at steady state operation with 3 out of 4 nozzles.
- A margin is required to account for aeration of liquid in the inner reservoir and for non-steady state flows and upsets.
- This design required a 40% turndown. To accommodate this, the bottom of half the notches are set higher (bottom of notch 150 mm above tray deck vs 50 mm for other half of the notches) to ensure an adequate liquid level in the inner reservoir at turndown for good flow distribution (similar to using orifices at two levels in a packed column liquid distributor).
- Notches are used in preference to holes. This is due to the high flowrate across the tray which would have required an enormous number of orifices.
- The higher the liquid height in the inner reservoir, the more even the liquid distribution through the notches, and the less sensitive the distribution is to the tray not being level.
- A high number of notches reduces the width of notch and increases weir loading which results in a high liquid velocity exiting a notch. High liquid velocities will increase the possibility of entraining vapour within the liquid falling from the inner reservoir to the annular sump ring which could result in poor vapour disengagement.
- High weir loads through the notches have a higher potential to generate froth. Thus the weir load is kept below 80 gpm/in.
- For the turndown case with all nozzles in operation, the minimum liquid level in the inner reservoir is kept above 6". This ensures good flow distribution to the annular sump ring compartments.
- The number, width and height of notches were optimised for liquid loading and better distribution.

- 100 mm wide notches spaced 50 mm apart are used. Two different heights of notches are used. The tall notches are located 50 mm from the floor of the inner reservoir deck and the short notches are located 150 mm from the floor of the inner reservoir deck. These different notch heights are used to cater for turndown case. There are 50 mm stiffening rings above the top of the notches.
- Notches to be field adjustable before any welding so that good alignment of the notch bottoms can be achieved.
- The notches were sized using the following rectangular notch equations in Perry's Handbook[4].

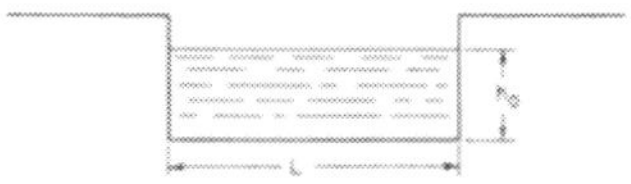

Figure 6. Flow through a rectangular weir for $h_o<L$

$$q = 0.415\,(L - 0.2h_0)\,h_0^{1.5}\sqrt{(2g)} \qquad (1)$$

[Ref.4, Eq. 10-39]

Flow through a Narrow rectangular notches for $h_o>L$

$$q = 0.386\,L\,h_0^{1.5}\sqrt{(2g)} \qquad (2)$$

[Ref.4, Eq. 10-40]

Where,
q = volume flow rate, m^3/s
L = crest length, m
h_0 = weir head, m
g = local acceleration due to gravity, m/s^2

CHIMNEY ARRANGEMENT

- The cross section and the number of chimneys was optimised to achieve
  - The required open area for the gas flow rate and
  - The required cross sectional area ratio to match the flow ratio requirement of the tray pass section in the distillation tray above to ensure the required vapour distribution to the tray above. (Chimney area for each section is proportional to open area of active panel above that section).
- The chimneys were arranged to minimize liquid flow obstruction and avoid significant liquid gradients in the inner reservoir. This results in a near constant liquid level on the inner reservoir and thus equal flow to each nozzle compartment.
- Circular chimneys were used as this provides less restriction to liquid flow when compared to rectangular or square chimneys. These are also more robust, especially when a liquid level is maintained on the external surfaces of the chimneys.
- A rectangular pitch is used for chimneys as this provides less restriction to liquid flow when compared to triangular pitch.
- No vapour chimneys are placed close to notches. This avoids flow restriction in front of the notches, allowing the liquid level to equalize in front of the notches (calming zone).

- Chimney tray is to be seal welded and water-tested for leakage.
- The chimney (vapour riser) height is to provide adequate margin above the maximum liquid level in the inner reservoir such that liquid will be prevented from overflowing the vapour risers for all foreseeable operating scenarios and upsets.
- The free area on the tray needs to recognise the potential aeration of the liquid flowing onto the chimney tray, as it may not be fully degassed in the downcomers from the active tray above.
- More riser area was provided for vapour flow and uplift resistance
- Chimneys height was set 150 mm above top of the notches upper stiffener rings.

OFF-CENTER DOWNCOMERS FROM TRAY ABOVE

- No Downcomer seal pans were used. The downcomers were submerged in the liquid of the inner reservoir. This was done due to concerns about the seal pan overflows generating froth on the chimney tray which may impact on the liquid heads on the tray. This design minimises the potential for liquid entrainment by vapours exiting the chimneys into the active tray above.
- Downcomers were sloped towards the centre to allow more chimney space on the inner reservoir and a more optimal chimney distribution on the tray.
- Flow path between the downcomer exits and shroud notches was to be kept as clear as possible.
- The design split each of the two downcomers into a pair with a 400 mm space in the middle between the pair to help equalize liquid level across the tray.
- The Downcomers have a 140 mm clearance to tray floor to maintain a liquid seal and provide adequate downcomer backup throughout the operating range.

CFD

To the best of our knowledge, our innovative improved design has not previously been applied. Therefore, it was necessary to audit the design using Computational Fluid Dynamics (CFD).

Computational Fluid Dynamics (CFD) was used in a parametric study to analyse the behaviour of the tray. The Gambit package was used for model construction and grid creation due to the need to control the cell count, grid quality, and grid spacing considering the size of the model and velocities. The grid used, contained about 3.6 million cells that were almost solely hexahedral. A Volume-of-Fluid (VOF) multiphase model with geometric reconstruction was used in Ansys Fluent 17.2 where the k-omega SST turbulence model was chosen. Transient simulations were run in parallel but still required considerable wall-clock time to complete. Time steps of 5 milliseconds were used and several seconds of simulated time were needed for the statistics of the chaotic flow to become relatively unchanging.

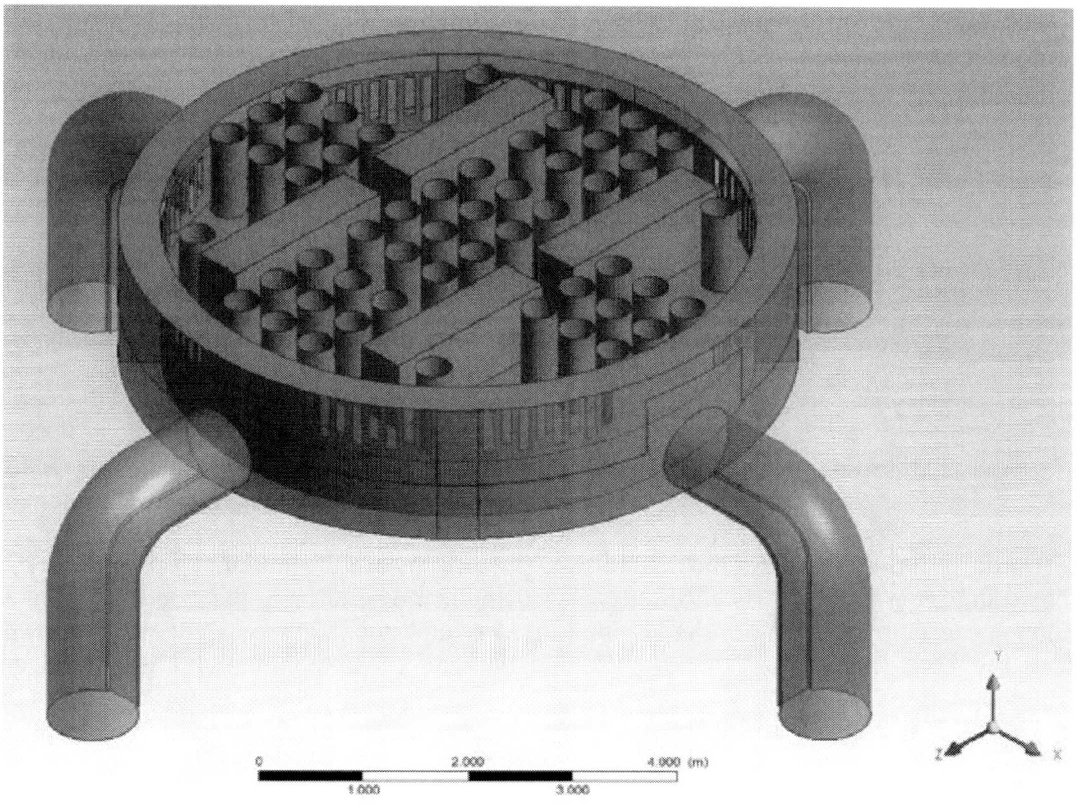

Figure 7: CFD Model of the Distributing Chimney Tray Design for the Crude Stabilizer

Depicted in Figure 7, the model included:

- The inner elevated tray deck with the vapor chimneys
- The downcomer sumps located on the elevated inner decks
- The notched annulus
- The segregated outer sump
- Reboiler feed piping

CFD was used to study tray tilt, inlet maldistribution, aeration, turndown, surge flow, and reboiler maintenance. The parametric study was split into three tranches of studies; namely, the ideal plate under normal operating conditions, non-ideal tray under upset conditions, and non-ideal tray at turndown.

The specific criteria used to analyse the results from the CFD were:

- Achieving equal distribution within 15% to all downcomers in service
- Preventing overflow of liquid into the chimneys
- Preventing excessive sloshing on the deck and in the sump which feeds the reboilers
- Preventing choking of the outlet nozzles resulting in back flow onto the tray

CFD of the ideal tray under normal operating conditions revealed that there was excessive sloshing in the annular sump, as shown in the image below, resulting from the liquid forces coming through the slots. Subsequently, the tray deck was lowered in the simulations which led to a marked improvement in the flow behaviour of the sump. All subsequent simulations used the lowered deck.

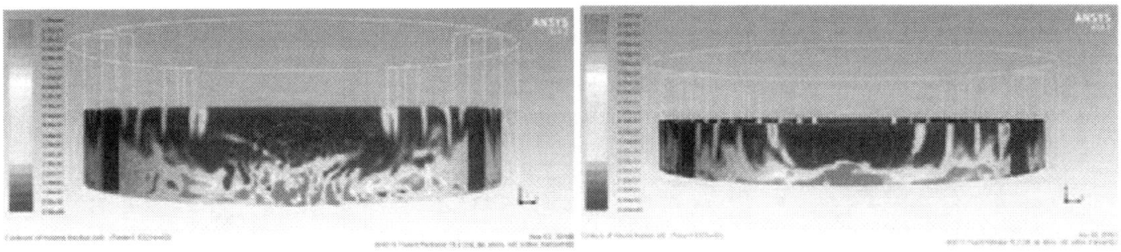

Figure 8. Sloshing in annular sump was excessive as designed (left) but much reduced by lowering the deck (right)

CFD allowed more in-depth understanding of the origins of the annular sump behavior. As shown in Figure 8, the levels of turbulence, sloshing, and aeration in the as-design annular sump is caused by the jets of liquid exiting the slots and impacting the walls. With a lower deck, the jets do not hit the exterior wall and have reduced momentum with less distance to fall as shown in Figure 9.

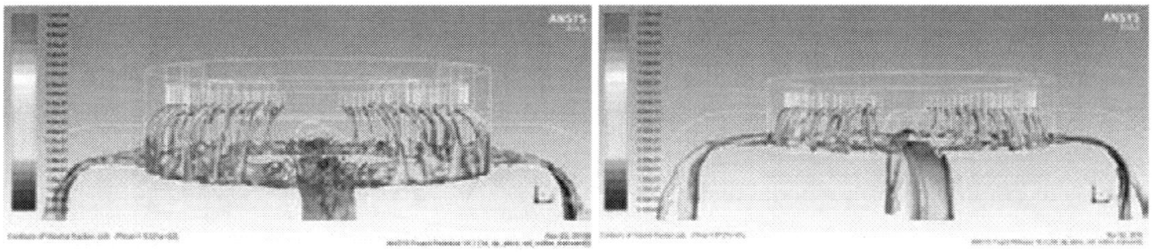

Figure 9. Liquid jets from the slots drive the behaviour in the annular sump with the as-designed case (left) having more sloshing compared to the lowered deck case (right).

CFD allowed understanding of the liquid height on the deck and degree of aeration throughout the parametric study which included varying the aeration in the incoming streams, tilting the unit, and varying throughput. Example images are shown in Figure 10 of the liquid level under ideal conditions (top). The middle and bottom images show the tray tilted with left side lower than right side as well as increased flow to the left downcomer and aeration in the downcomers (middle shows liquid/gas interface and bottom shows relative vertical coordinate of liquid/gas interface).

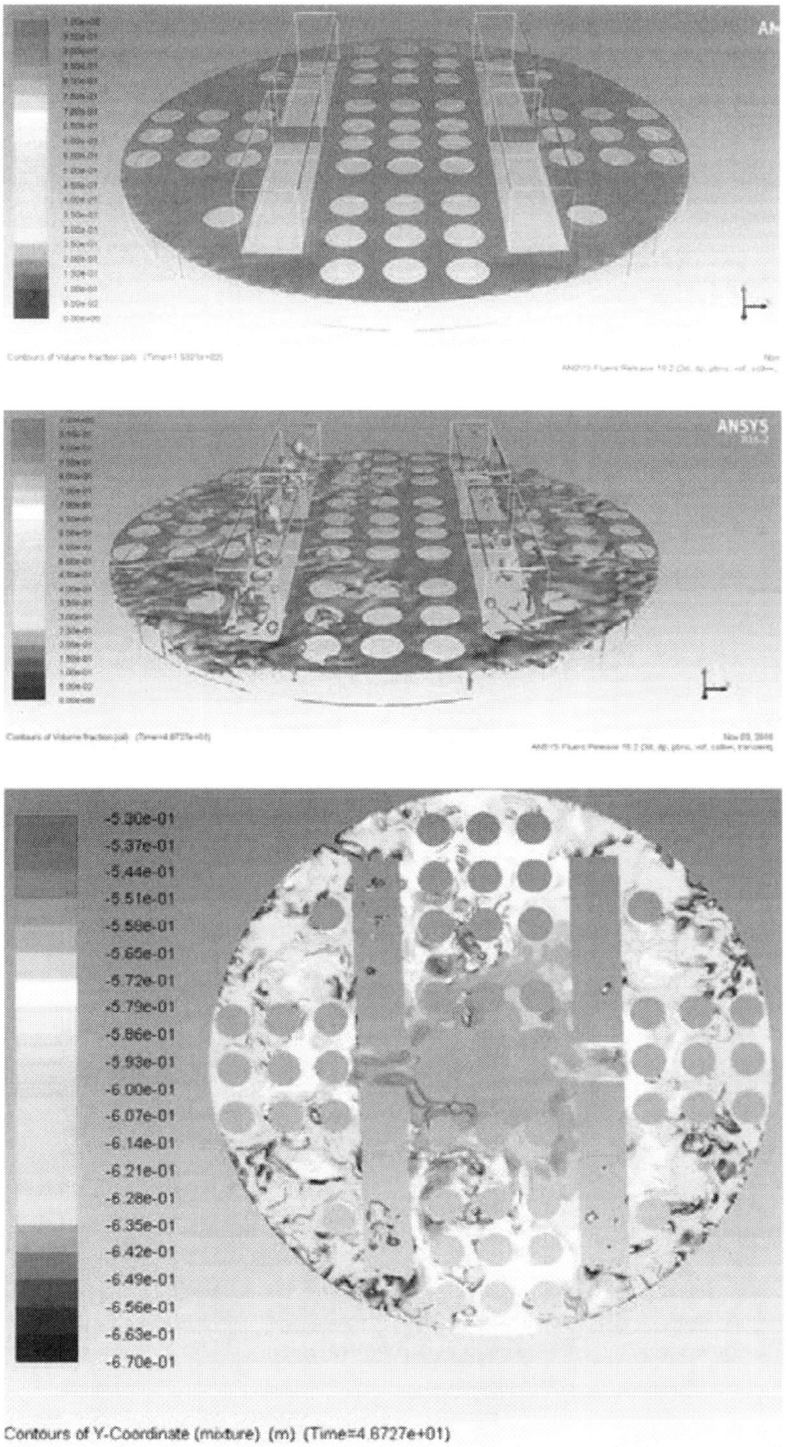

Figure 10. Contours on the Stabiliser draw tray.

The tray performed very well under non-ideal conditions. Shown below in Figure 11 is the relative performance of the tray when subjected to the variations listed above. Performance was measured by variance from equal flow at all operating nozzles. The only instance where the tray exceeded the 15% limits was during turndown. In the turndown situation, the liquid level on the tray is so low that the short slots are not being used. The variance relative to the lower overall flow rate exceeded the desired 15% cut-off with all nozzles operating, whereas the absolute variance (kg/s) in flow rates between nozzles was actually less than the 15% limit when flowing at design rates. Performance improved during turndown when maintenance shut one of the nozzles, thereby increasing flow to the other nozzles and making the absolute variance (kg/s) fall within the 15% relative variance acceptance levels.

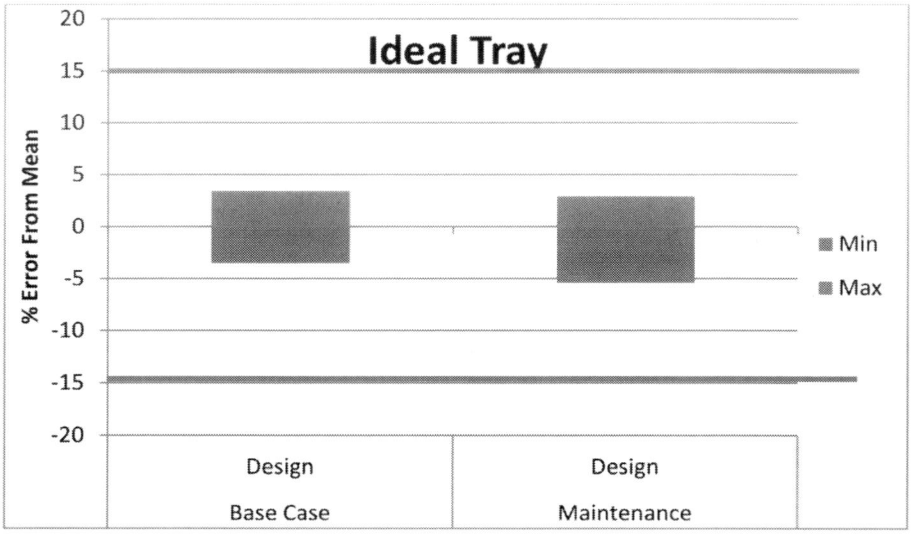

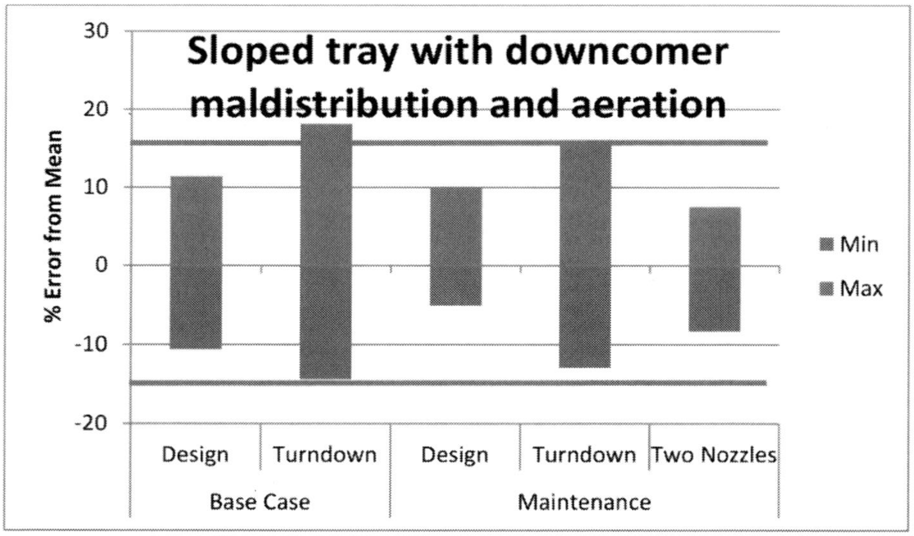

Figure 11. relative performance of Stabiliser draw tray.

Full Scale Water Testing of the Reboiler Draw Trays

As the Crude Stabilizer and Gasoline Splitter are key to the performance of the facility it was decided to construct and test a full scale prototype of the proposed tray design. Figure 12 shows 3D-Model and Figure 13 shows actual photographs of the test facility for the MOTR. This was done for both towers at the Sulzer fabrication facility in Tulsa, Oklahoma, USA.

The following tests were carried out,

- With all nozzles in operation at 100% of design flowrate
- One nozzle taken out of service at 100% of design flowrate
- Turndown operation at 40% of design flowrate with one and 2 nozzles taken out of operation
- 60/40 maldistribution in the downcomers from trays above at 100% of design flowrate
- 115% of the design flow all nozzles in operation

Due to the constraints of the test facility and tray construction it was not possible to test the distribution with induced tilt on the tray.

Results showed that the maximum maldistribution is less than 2% for the tests, which was similar to the CFD results. By reducing the fluid density in the model, the results could be extrapolated to mimic the aerated liquid from the downcomers, and these too were similar to CFD results.

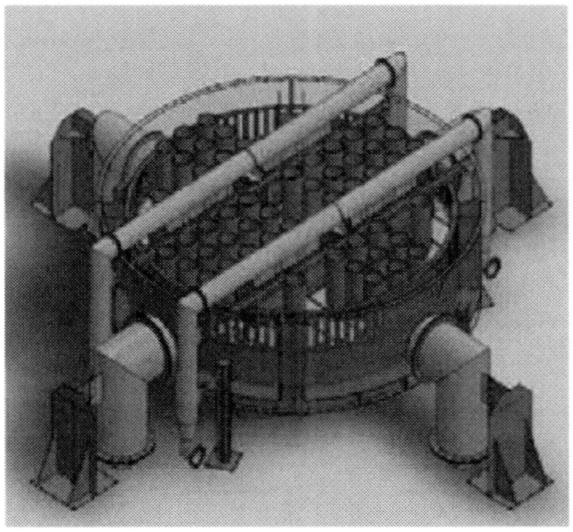

Figure 12. 3D-Model of full scale testing of the Crude Stabilizer Reboiler Draw Tray

Figure 13. Photographs: Full Scale Water Testing

Conclusions

In conclusion, this MOTR splitter chimney tray design was demonstrated using CFD and full-scale water test to -

1. Ensure equal distribution of liquid to multiple reboiler draw lines over entire operating range including turndown.
2. Accommodate equal redistribution of liquid to active reboilers when one or more reboilers are isolated for maintenance. This means that the reboiler design does not require additional margin to overcome maldistribution and the design rate can be achieved.
3. Achieve complete decoupling between the distributing element and the reboiler circuits, avoiding swings and transients.
4. Provided that the annular sump is segmented to provide an equal weir secant and notch cross sectional area, it can ensure equal reboiler feed for a range of layout arrangements for multiple reboilers as the flow is evenly distributed at the source.
5. The impact of tilt on the draw tray is minimised.
6. Reasonable maldistribution from the 4-pass tray above has no impact on reboiler feed distribution.

7. Although not the purpose of this design it is probable that by careful design of the notch cross-sectional area the principles can be used to preferentially feed one reboiler system compared to another.
8. Results do show that for turndown scenarios, closing additional nozzles will improve distribution performance.

References

1. The FRI Design Practices Committee, "Reboiler Circuits for Trayed Columns", Chemical Engineering, January 2011.

2. Jacobs, J. K., "Reboiler Selection Simplified," Hydrocarbon Processing and Petroleum Refiner 40:70 (1961), p. 190.

3. Kister, H. Z.,"Reboiler Circuit Debottleneck with No Hardware Changes", Chem. Eng., August, p. 30, 2017.

4. Boyce, M. P., V. H. Edwards, T. W. Cowley, T. Fan, H. D. Kaiser, W. B. Geyer, D. Nadel, L. Skoda, S. Testone, and K. L. Walter, "Transport and Storage of Fluids", Section 10 , in R. H. Perry and D. Green "Chemical Engineers' Handbook", 8th Ed., 2008.

5. Kister, H. Z., "Distillation Troubleshooting", Wiley-Interscience, 2006.

546234 Identifying Turndown Limits for Vertical Tubeside Thermosiphon Reboilers

Operating thermosiphon reboilers at less than design capacity is often necessary, but the lack of knowledge as to how far a unit can be turned down can cause problems. At HTRI, we have recently completed a set of experiments in a thermosiphon flow loop to answer the question of turndown limits. A vertical tubeside thermosiphon reboiler (VTTR) was attached to a separator. Steam supplied to the shellside of the VTTR heated up the tubeside fluid, which came in subcooled and exited two-phase. The density difference between the two-phase fluid in the VTTR tubes and the single-phase fluid inside the column drove the circulation. By controlling the column pressure and systematically lowering the mean temperature difference (MTD) between the steam and the process fluid, the unit was turned down gradually. Below a certain MTD or average heat flux, we observed an end to liquid recirculating back into the column, marking the turndown limit. Based on tests with pure fluids and wide boiling range mixtures, very different turndown mechanisms were observed, indicating different underlying mechanisms responsible for unit shutdown:

- Pure fluid tests indicated two distinct behaviors: tests with a light hydrocarbon (n-pentane) indicated that the unit was operational even with an MTD as low as one Deg-C, the lowest MTD we could attain experimentally. Tests with water and p-xylene indicated that below a minimum MTD or heat flux, the vapor generation rate matched the circulation rate, pointing to a pool boiling type mechanism with all vapor outlet. For pure fluids or fluids with low boiling range (< 5 Deg-C), the only known shutdown criterion is based on the end of onset of nucleate boiling (ONB). Our data clearly indicates that for some pure fluids, the ONB criterion is insufficient.
- Tests with wide boiling range (> 30 Deg-C) mixtures indicated that below a certain minimum heat flux, the circulation abruptly ended. We postulate that just before shutdown, the low heat flux and low circulation rate led to lights boiling off preferentially, resulting in accumulation of the heavy component. This led to a rapid lowering of the effective MTD, causing the unit to shut down.

While this research is not complete, our observations and analysis clearly indicate the need to develop better criteria for turndown limits than those currently available. HTRI is currently working on developing guidelines to help the design, operation and troubleshooting of thermosiphons. Additional experimental work with other fluid mixtures and different turndown schemes (e.g. flooding the VTTR shellside to reduce the effective heat transfer area) are also being considered.

551758 UOP High Cond^{TM} Tubes Debottleneck Fractionation Research Inc. (FRI) Condenser

Lack of condensing capabilities limit many fractionation units throughout the refining, petrochemical, and gas processing industries. This issue can be exacerbated, especially during the summer months, when cooling water temperatures rise and operators are forced to increase the column pressure to maintain the required temperature difference across the condenser. Increasing the operating pressure reduces the relative volatility between the key components, often resulting in higher energy consumption and lower production.

Fractionation Research Inc., located in Stillwater, Oklahoma and better known as FRI, is a non-profit consortium supported by memberships which include many of the largest petroleum, chemical, and engineering companies in the world. Founded the 1950's, FRI conducts commercial scale distillation testing with various hydrocarbon systems at pressures from deep vacuum to 500 psia.

Over many years, FRI encountered the issue described above, with a significant condenser bottleneck limiting the ability to fully test the newest generation distillation internals in their 4 foot and 8 foot diameter test columns. With an original design duty of 15 MMBTU/hr, the low pressure condenser could not generate enough reflux to flood the newest high capacity devices.

In late 2017, FRI replaced their existing bare tubes condenser with UOP High Cond tubes condenser, providing a 67 percent increase in duty while maintaining the existing foundation, piping, and plot space. UOP High Cond tubes provide significant improvements in the heat transfer coefficient in horizontal cooling water condensers over bare tubes due to a proprietary surface enhancement.

This paper will discuss the UOP High Cond Tube performance, advantages and applications. The focus of the paper will be primarily on the design of experiments, the performance, and the results of FRI's High Cond tube condenser. FRI, as a research facility, measured and quantified all performance variables by adding instrumentation not typical for commercial process equipment. The UOP High Cond Tubes condenser performance has exceeded FRI's expectations, and has successfully eliminated their summertime cooling water limitations, and has enabled the FRI experimental unit to operate at lower column pressures than were previously achievable.

Overcoming Seasonal Cooling Problems in Distillation Column Overheads - Case Studies

AIChE Distillation Symposium, "New Insights into Reboilers and Condensers"
Session 6, Wednesday, April 3, 2019
"Honoring the Contributions of Greg Shinsky: A Jubilee of Advances in Distillation Control"

Umesh Mathur, P.E. (umesh.mathur@aggreko.com), and
Bryan Markland (bryan.markland@aggreko.com)
Aggreko Process Services
3732 Magnolia Street, Pearland, TX 77584

Contents

1. BACKGROUND

Some of the most ubiquitous and, in our opinion, most serious problems confronting distillation operators are the results of seasonal or diurnal variations in ambient air and cooling water (CW) temperatures. Generally, distillation column overhead cooling systems are designed for some sort of average summertime CW or air temperature. These are routinely exceeded, according to seasonal weather data, for extended periods during the peak summer months. Also, over time, column throughputs have been increased as units were debottlenecked or when other process changes were made. On occasion, purity specifications may change, inexorably becoming more stringent.

As Kister has very ably and clearly pointed out[1], column pressure control is the single most important variable when it comes to proper operation of day-to-day distillation systems. We find that this issue is poorly understood in many major refining and petrochemical sites. Unfortunately, there are a number of confounding misconceptions regarding the "benefits" of allowing column pressure to float with ambient air or CW temperatures. In this discussion, we focus on some of these issues and present our experiences in several typical situations. The serious effects of permanent increases in tower operating pressure require cutbacks in column feed rates. This issue is discussed further below.

In this paper, we show examples of how it is feasible to restore column overheads cooling capacities to winter-like conditions by installing temporary, or seasonal, supplemental cooling solutions. The key is to be able to add such capabilities without intrusive hot taps or any interruption or disturbance to unit operations. We also discuss some of these techniques that have been proven successful in a large number of past applications in the petroleum refining, petrochemical, and complex chemicals industries.

The benefits of such seasonal enhancements in summertime operations can be extremely valuable, especially if inadequate capacity in downstream columns in a unit are cascading back to the front-end of a refinery or major petrochemical facility. An example would be cutbacks at a debutanizer in a fluid catalytic cracking unit (FCCU) unsats gas plant that causes not only reduced charge to the FCCU but, ultimately, the crude and vacuum units at the refinery. Such losses could easily amount to millions of dollars a month or more. Preventing such losses using supplemental cooling reverses these losses with a very high benefit to cost ratio.

2. DISCUSSION

We first elaborate on some of the types of problems caused by inadequate distillation overheads cooling. These include:

2.1 Inadequate Column Overhead Pressure Control

Some columns with total condensers allow the pressure to float with ambient air or CW temperature, as they have no effective pressure control strategy at all. This lamentable situation leads to a diurnal cycling of tower pressure that resembles a sine wave. Such cycling in pressure constitutes a serious disturbance to column operation and unleashes composition transients throughout the column. In superfractionators and other high-purity distillation columns especially, these composition transients can have a time-to-steady-state of a day or even more. As a result, the operator begins to chase the composition at both ends of the column, sometimes making large simultaneous moves in both overhead reflux and bottoms reboiler duty. These moves have their own effects on column dynamics and product purity control can therefore become quite problematic.

In a paper presented at a previous AIChE Distillation Symposium in 2007[2], it was shown that first-principles dynamic simulation of distillation columns can be extremely valuable in understanding these transients. An additional benefit of such models is that they can be used to minimize, if not eliminate entirely, costly and disruptive plant step tests of columns to obtain the "dynamic matrix" of curves that is required for optimal, multivariable control. The main point is that the composition disturbances can be long-lived and their time period can sometimes be of the order of a day or even longer. Thus, the column could still be reacting to overhead cooling problems caused by reflux temperature / flow rate or pressure cycling on a given day but get hit with the same cycle at the beginning of the next day. This is one reason why some, if not many, real-world distillation columns never see a true steady-state.

Other distillation columns may have faster dynamics but are burdened with an improper regulatory pressure control scheme that makes it difficult, if not impossible, to maintain the desired product purity set-points adequately. Such inadequacies can easily cause severe periodic fluctuations in product quality, especially when key-component relative volatilities are influenced strongly by tower pressure.

In still other cases, the condenser effluent is a two-phase mixture, and the column may make both vapor and liquid distillate products. Examples include refinery fractionators in crude units, FCCUs, hydrocrackers and hydrotreaters, delayed cokers, etc. In these units, there is a wet gas compressor to raise the pressure of the vapor distillate sufficiently to enable entry to the associated gas plant. Inadequate cooling of the condenser in the summer increases the flow of uncondensed vapors to the wet gas compressor and may easily threaten to exceed the maximum intake capacity of the machine. The operator would then have little choice but to cut back unit throughput.

Other consequences may also be attached to such overhead cooling limitations: the fractionator liquid reflux stream flow might become inadequate to maintain the separation between adjacent side-draws in the fractionator. In response, the operator might take other actions: for example, he might change feed preheat, pumparound duties or adjust sidestripper steam injections or reboiler duties. All such actions can affect unit throughput. Also, keeping product qualities on specification may mandate significant downgrading of valuable products to less valuable ones. Finding the proper control strategy can often be tricky and not self-evident in such situations. While multivariable controls can be effective in mediating these effects, they too are often hampered by lack of reliable on-line estimates (e.g., inferential calculations) of sidestream qualities[3].

We do not wish to belabor these operating issues further, but would say that the best solution would be to ***solve the problem at source***: provide supplemental cooling in the column overhead cooling system. To be acceptable in a complex operating environment, these additions need to be made in a graceful, non-intrusive, and reliable way that restores operation to winter-like conditions without disturbing the process or requiring a shutdown. These techniques are described in later sections of this paper.

2.2 Throughput Loss Caused by Inadequate Overheads Cooling.

During the summer, ambient air temperature can be 40-50°F higher than in the winter, and CW temperature is often is higher by 30°F or more. Inadequate summertime cooling for distillation column overheads has a number of highly deleterious consequences:

- If the column has a vapor distillate, the flow of this vapor rises dramatically as overhead coolant temperature increases. When the vapor is sent to a compressor (e.g., when feeding a gas plant in a refinery main fractionator column), the compressor inlet volumetric capacity limits column feed rate.
- The reduced temperature driving force in the overhead condensers requires column pressure set-points to be raised considerably.
- At higher pressures, the column separation suffers as a result of reduced relative volatility between the key components.
- In an effort to maintain separation efficiency, the operator feels obliged to increase reflux and reboiler duties. This brings the column operation even closer to its flooding limit, with a corresponding increase in liquid entrainment and a further reduction in separation efficiency.
- Eventually, this cascade of events forces operators to decrease column feed rate, often by cutting the throughput to the entire unit.
- Cutbacks of 3-7% are not uncommon, since process control schemes can only push a unit to the operating limits imposed by ambient conditions.

The economic consequences of cutbacks in unit throughput are extremely serious, especially when units are required to be running at their capacity limits. Conversely, avoiding such cutbacks using well-designed and engineered solutions, on a temporary basis, can restore profitability. Such projects can and should be implemented reliably on a very fast track (typically days or weeks) and adhere to industry standards for workplace and process safety, and also environmental compliance.

2.3 Product Quality Violations

Fluctuations in tower pressures, feed rates, or reflux rates invariably set off composition transients that require significant countervailing moves for maintaining product quality at both ends of a tower. Often, operators intervene in ways that, eventually, lead to potential violations of product quality specifications, or throughput reductions, or both. One tactic, highly inadvisable but often practiced routinely, is to make overheads product of a far better quality than specification so as to be able to absorb occasional violations, especially for high-purity specifications. Such giveaways lead to (a) lost production when column capacity is near its limit, or (b) excessive utility consumption, which can be quite costly especially in columns using refrigerated overheads cooling.

These points have been noted in the distillation literature repeatedly. However, the traditional approach is to focus on improved regulatory and advanced control schemes to minimize production losses and product quality excursions. In this discussion, we show examples where improving overhead cooling to restore separation efficiency and throughput to winter-like conditions has proven to be a superior alternative from an economic perspective.

There are several innovative and highly profitable ways to provide direct supplemental (or replacement) cooling of column overheads during the summer months. These have been field-proven over many years in a variety of process technologies throughout the refining and petrochemical industries. The goal of these approaches is to move column operation back to winter-like conditions and restore production and product quality reliably to desired levels. These temporary solutions require application of sound engineering principles, using the proper process simulation, phase equilibrium thermodynamics, and equipment rating techniques fully commensurate with the highest industry standards.

Capital appropriations are often hard to justify for seasonal cooling limitations. There is a paramount need for reliable solutions that can be engineered, installed, and commissioned in days or weeks to capitalize on short-term product margins. This situation is best handled using industrial-grade rental equipment such as heat exchangers, refrigerated chillers, cooling towers, etc. The key is to be able to insert such equipment into a running column in a way that does not require a shutdown and causes no disruption or disturbance in column operations. Such seasonal projects have been proven in numerous successful industrial applications.

We present below examples of such engineered solutions pertinent to the refining and petrochemical industries that have yielded benefit/cost ratios of ranging from 4:1 to 10:1 or even better when complex-wide production rates were increased, sometimes dramatically. Best of all, the operating problems mentioned above are eliminated, or at least ameliorated significantly, using these approaches.

3. COLUMN OVERHEAD COOLING TECHNIQUES

Several methods can be used to provide supplemental or replacement cooling for distillation column overheads. We list these below and then discuss them individually:

- Cool overhead reflux
- Cool overhead accumulator using reflux pump(s)
- Replace condenser CW with dedicated cooling tower loop
- Replace condenser CW with dedicated refrigerated coolant loop

The main point to understand is that the column itself benefits from such overheads cooling strategies, no matter how they are, in fact, implemented. The proper choice depends solely on the extent of supplemental cooling desired and the feasibility of each of these approaches in a given instance.

3.1 Column Overhead Cooling Options

Often, we find that the cooling towers in a major refinery or petrochemical plant were designed and built many decades ago and now handle far greater CW circulation rates than design. The result of such overloading is that the CW supply temperature back to the plant has increased steadily over time. Increases of 10°F or even more in excess of design are not uncommon. This reduces the temperature driving force for all cooling applications and has a highly deleterious impact on the cooling of many major units, especially column overhead condensers.

In cooling a distillation column, it really doesn't matter what mechanism is chosen to achieve the desired result: the overhead reflux must return to the column at a considerably lower temperature than currently.

Such supplemental reflux subcooling is highly beneficial:

1. By subcooling the external reflux, it increases internal reflux liquid flow below the top tray of the column, thereby improving fractionation or enabling increased throughput
2. By condensing part of overhead vapor, it reduces the duty on an overloaded overhead compression system.
3. It enables reducing column operating pressure to provide improved separation.

Each of these enhancements can also enable increased unit throughput. Aggreko has developed several methods to accomplish these objectives. Each method has its own advantages and the best choice is made after a site-specific study in every case. It must be emphasized that such projects require a very high degree of engineering and technical skill in field implementation to ensure safety, robustness, reliability, and flawless execution.

The supplemental equipment must also be of the highest industrial quality, fully compliant with applicable standards and client specifications. Aggreko offers a large fleet of well-maintained heat exchangers, refrigerated chillers, electrical power generators (both diesel- and gas-fired), cooling towers, pumps, and ancillary equipment. All heat exchange equipment is cleaned thoroughly and professionally prior to each new project to eliminate the possibility of contamination.

Such equipment is used not only for distillation column overhead cooling but also many other applications such as increasing oxygen in the air supplied to FCCU regenerators, product rundown cooling, chilling absorber lean oils for improving recovery of LPG components from fuel gas, cooling the lean amine stream to absorbers to improve acid gas recovery, and many others.

We now describe the various cooling options and configurations specifically for distillation column overheads. These rental solutions have been applied very successfully over the years, yielding highly attractive rates of return, ranging anywhere from 400% to 2,000%.

These procedures enable the equipment to be inserted safely and without any interruption in column operation. No disruption is caused and there is certainly no requirement for a maintenance outage to make this installation. Specifically, we adhere to all client requirements and participate in process safety reviews and management of change (MOC) procedures to ensure process and personnel safety. There are many precautions and procedural safeguards to be respected, but we do not dwell on them here.

3.2 Cool Overhead Reflux

The simplest and most direct way of providing additional cooling for a column's overhead system is to cool the reflux stream significantly below the temperature of the liquid in the reflux drum, Figure 1. This liquid exits the condenser, a few degrees above the condenser coolant supply temperature. When CW is used, additional cooling generally requires use of heat exchange equipment with a chilled circulating coolant. The coolant is chilled using mechanical refrigeration in a closed-loop.

The key question to be addressed is: how do you insert such equipment in a way that does not disrupt unit operations or require a maintenance shutdown? A unique methodology was developed by Aggreko for achieving precisely this capability that is described in U.S. Patent US9134064B2[4]. The basic idea is to:

1. Pre-fabricate spool pieces that are connected between the isolation block valves on each side of the reflux flow control valve.
2. Divert the reflux flow temporarily through the control valve bypass line
3. Close both control valve isolation block valves
4. Remove the control valve
5. Tie the prefabricated spool pieces from the upstream bock valve to the heat exchange equipment inlet (relocating the control valve into this line) and from the exchanger outlet to the downstream block valve
6. Connect the coolant circuit piping for the exchanger
7. Tie the mechanical refrigeration equipment and circulating pump into the coolant circuit
8. Charge the coolant circuit with water or a glycol-water solution, as required
9. Commission the chiller

10. Gradually close the bypass valve and open the block valves to divert the reflux flow through the temporary loop where it is chilled to the desired outlet temperature and returned to the column.

Aggreko executes these steps in a way that adheres to all engineering and operational requirements from our clients. This procedure has been used very successfully in numerous of past projects.

3.3 Cool overhead accumulator using reflux pump(s)

In a manner very similar to (3.2) above, it may sometimes be desirable to cool not only the reflux but the overhead distillate vapor and liquid products also, Figure 2. This would, for example, be desirable if the vapor distillate is being sent to an overloaded wet gas compressor in a refinery crude unit, FCCU, delayed coker or other similar process that feeds a gas plant for recovering propane and heavier components from the fuel gas product.

To do this, the stream leaving the supplemental chilled exchanger would be split, part of the flow being directed to the column as reflux. The other part would flow either into the overhead condenser outlet line to the reflux drum (or into the drum itself in a way that ensures good vapor-liquid mixing). Either way, heavier components that were flashing off to the vapor distillate would be condensed into the liquid distillate, thereby reducing wet gas compressor load.

The supplemental cooling load in this case would be higher than for (3.2) above because the distillated vapor and liquid products are also being cooled.

3.4 Replace Condenser Cooling Water With Dedicated Cooling Tower Loop

In some instances, it makes the most sense to replace the cooling system for all or some portion of the column overhead condensers entirely, Figure 3. Alternatively, when some such exchangers are being removed for cleaning, replacement exchangers could be installed (in parallel) and these (and perhaps all the other exchangers) could be cooled using one or more supplemental cooling towers in a dedicated CW loop.

Several benefits accrue from using our cooling towers:

1. The delivered CW is 4-5°F above the wet bulb temperature; generally, this is significantly lower than the plant's indigenous CW.
2. The amount of CW can be much greater than that available locally, especially for some units that are located at the end of a long circuit.
3. A dedicated CW loop can enable on-line chemical cleaning to unclog heavily fouled or even partially blocked exchanger tubes.

For example, in refinery vacuum distillation units, the column overhead pressure is quite low, often 50-70[torr]. Pipe sizes are very large and, consequently, pipe lengths to the condensers are minimized. This requires placing condensers at a high elevation, sometimes over 60[ft] above ground level. This design often restricts the CW flow to these condensers and therefore the CW outlet temperature rises to a level that promotes accelerated deposition of salts in the exchanger tubes.

In past projects, we have provided dedicated heat exchangers, running in closed loop mode with our cooling towers or chillers, to greatly increase flow while also reducing CW temperature to maximize cooling. This enables achieving a lower pressure in the column's flash zone and reduces the loss of valuable gas oils (normally fed to the FCCU) to the bottoms asphalt stream.

3.5 Replace Condenser Cooling Water With Dedicated Refrigerated Coolant Loop

In a manner similar to Section (3.3), it can sometimes be desirable to replace the existing CW supply to an exchanger with a dedicated circulating refrigerated chilled water loop to improve heat transfer and eliminate production bottlenecks, Figure 4.

In some prior projects, we have used this method for cooling refinery vacuum unit overheads that reduced column flash zone pressure significantly, increasing the flash vapor flow and thus improving the quality of the asphalt used for road surfacing. In one instance, the fouling of the overhead exchanger was so pronounced that the refinery was contemplating an unscheduled maintenance outage to clean the affected exchanger(s). This project was designed to improve the vacuum system performance but, because a closed-loop chilled water circuit was used, the refinery was also able to perform on-line chemical cleaning to unclog the fouled exchanger tubes. This improved column operation substantially and also enabled the refinery to continue routine operations until the next scheduled maintenance outage. The savings from avoiding an unplanned outage were enormous, as the upstream crude unit and several downstream units such as the FCC also stayed on-line throughout.

4. EXAMPLES

4.1 Refinery Crude Unit Main Fractionator

Our first example shows the effect of overhead cooling water temperature on column distillate vapor flow rate. Figure 1 below shows a crude distillation column with one pumparound and four side-strippers. Uncondensed overhead vapor flows to a compressor that feeds the gas plant to recover propane and heavier components from fuel gas.

The column has 40 trays plus an overhead condenser and reflux drum. The column is modeled using 27 theoretical stages. In Figure 1, the actual trays are shown alongside theoretical stages in parentheses. Uncondensed vapor distillate goes to a two-stage, intercooled wet gas compressor in the gas plant, while the liquid distillate (light naphtha) is pumped to the gas plant.

The four side-strippers are for heavy naphtha, kerosene, diesel, and light gas oil. The heavy naphtha stripper is reboiled, while the other three use stripping steam. Steam is also used to strip the column bottom product (long residue).

The following parameters are fixed for this simulation:

1. Pressures at column condenser drum, top tray, and bottom tray
2. The column feed rate, temperature, pressure, and composition leaving the feed preheat furnace
3. Overhead reflux flow (BPD
4. All stripping steam temperatures, pressures, and flow rates
5. The heavy naphtha side-stripper reboiler duty
6. The pumparound flow rate (BPD) and heat duty
7. Product flow rates (BPD) of light naphtha, heavy naphtha, kerosene, gasoil
8. Light naphtha ASTM D-86 distillation temperature
9. Condenser CW flow rate (maximum, fixed)
10. CW supply temperature
11. Condenser pressure drop on tube and shell sides
12. Condenser U*A (product of heat transfer coefficient and surface area)

The calculated condenser outlet temperature on the process side is used to set the reflux drum temperature. As the CW supply temperature increases, the condenser outlet temperature increases. This results in an increase in the inlet volumetric gas flow (actual ft^3/minute) to the wet gas compressor.

Several cases were run for CW supply temperature while all other simulation inputs (including tower pressure) were held fixed. The effect on the wet gas compressor inlet flow is shown below in Figure 2.

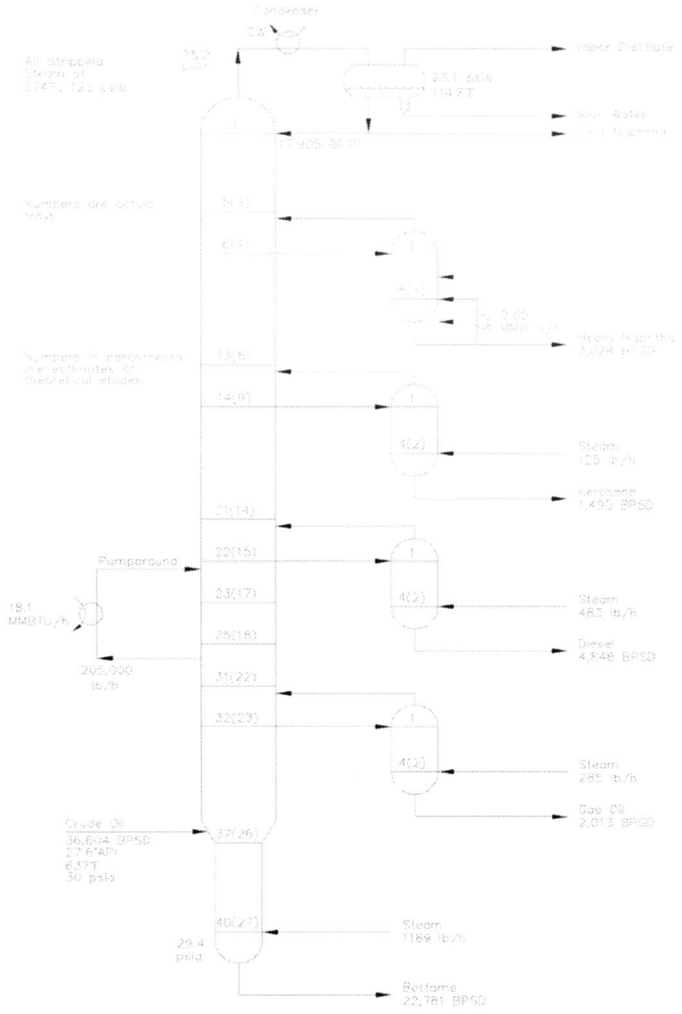

Configuration and conditions for the simulation of the atmospheric tower of crude unit.

FIGURE 1: SCHEMATIC OF CRUDE UNIT MAIN FRACTIONATOR

Umesh Mathur and Bryan Markland, Aggreko Process Services, Pearland, TX

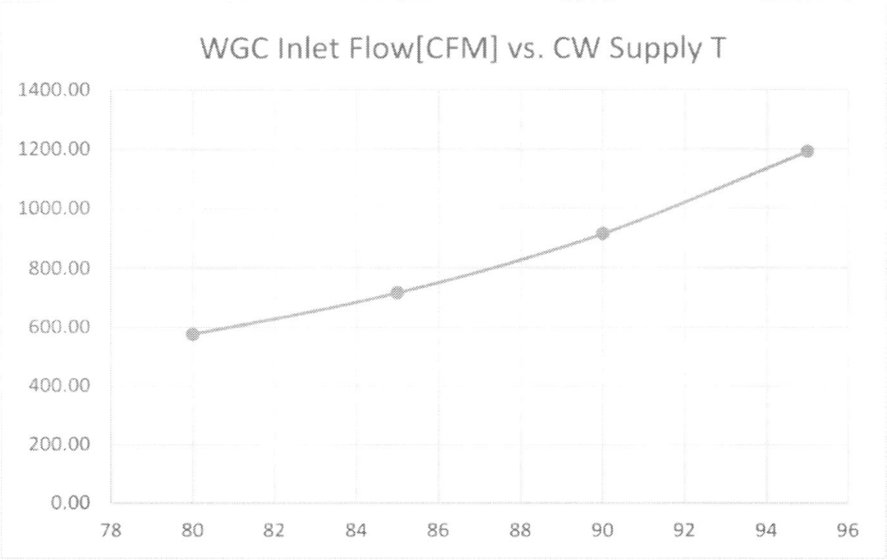

FIGURE 2: OVERHEAD VAPOR vs. COOLING WATER SUPPLY TEMPERATURE

This example shows clearly that the inlet gas flow to the wet gas compressor varies almost linearly with CW supply temperature. In past projects, lowering CW supply temperature from 94°F to 86°F, using dedicated cooling tower(s) for overhead condenser CW, enabled reducing the inlet gas load on the compressor by about one-third or nearly 33%.

Over time, the normal refinery CW temperature in most older plants increases over a period of time as units are debottlenecked and new units are added. Cooling towers continue to reject the increased heat load to the atmospheric air but the CW return temperature increases steadily with heat load. We have seen cases where the return temperature has increased by over 10°F, compared to design, over several decades.

The effect of increased CW supply temperature on process cooling throughout the refinery can be quite dramatic. To alleviate such problems, dedicated cooling towers have been used most effectively to supplant the normal refinery CW in a wide variety of localized cooling applications.

Raising crude tower operating pressure to reduce compressor load is not a desirable option because this also raises the feed preheat furnace outlet pressure. Generally, the heater outlet temperature is fixed at a limit to minimize thermal cracking of the feed and tube coking. Accordingly, increased tower pressure reduces the feed flash in the fractionator flash zone and reduces atmospheric gas oil production. This, in turn, increased the load on the vacuum tower feed preheat furnace and the vacuum tower loading. These phenomena, taken together, reduce vapor traffic in the main column, reduce reflux flow, worsen fractionation between the side-cuts (leading to excessive side-product downgrading), worsen vacuum tower performance, and, ultimately, hurt the economics of the operation.

Similarly, in an FCC, increased fractionator overhead pressure increases reactor pressure and coke make, reducing valuable products while also degrading main fractionator column performance.

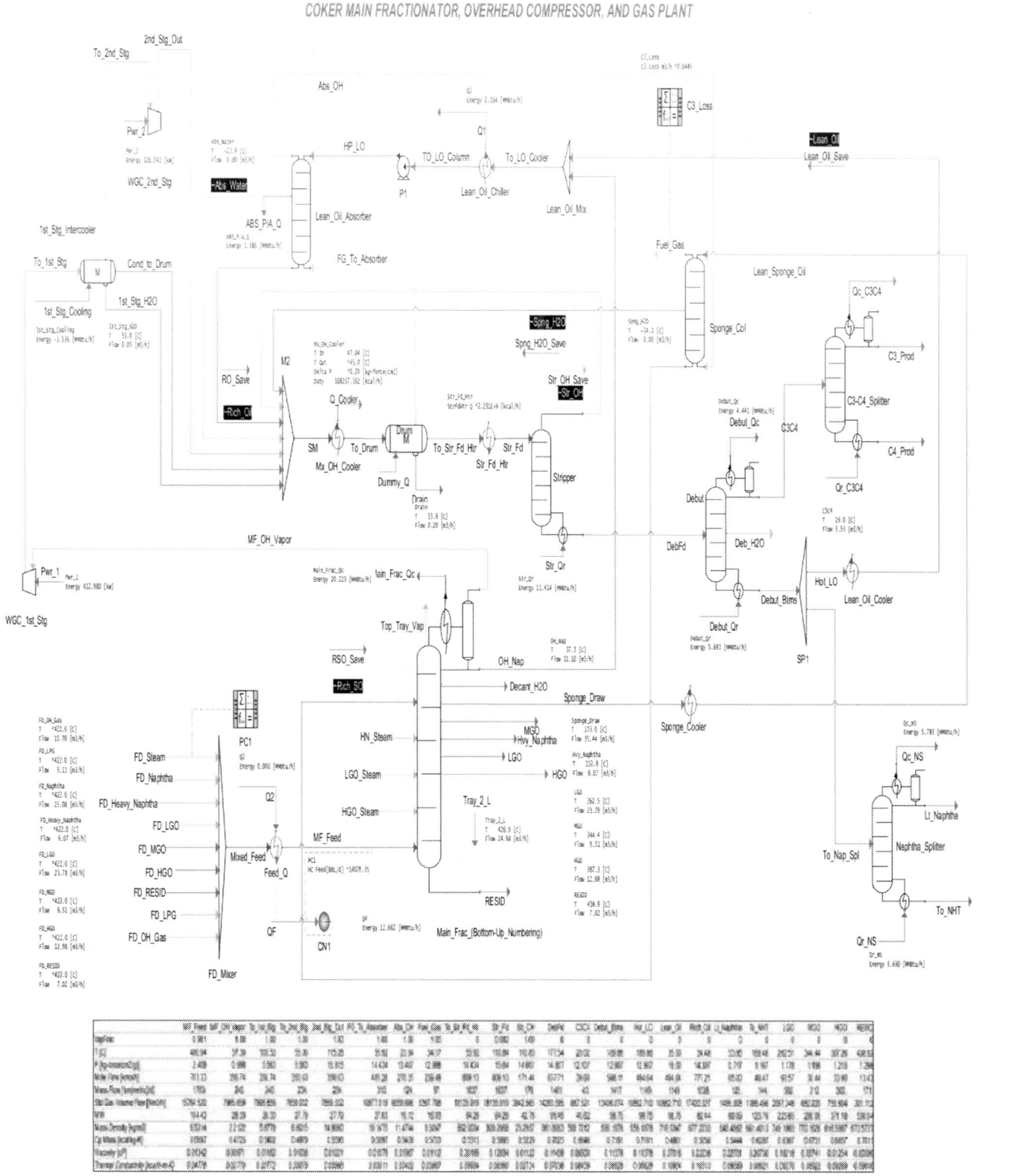

FIGURE 3: DELAYED COKER MAIN FRACTIONATOR, WET GAS COMPRESSOR, AND GAS PLANT

Umesh Mathur and Bryan Markland, Aggreko Process Services, Pearland, TX

4.2 Delayed Coker Main Fractionator, Wet Gas Compressor, and Gas Plant

This is a more elaborate example for showing the inter-connected problems resulting from separation system capacity limits in major petroleum refinery units. Figure 3 shows a process simulation diagram for a delayed coker main fractionator, overhead wet gas compressor, and gas plant for recovering C3s and heavier components. The separation columns in this unit are:

1. Main Fractionator (with pumparounds and side-strippers)
2. Lean Oil Absorber
3. Stripper
4. Sponge Oil Column
5. Debutanizer
6. C3/C4 Splitter, and
7. Naphtha Splitter

In addition to supplemental cooling of the Main Fractionator overheads, improved cooling of the lean oil to the absorber column can lead to dramatic improvements in the recovery of the more valuable C_3 and heavier fractions from fuel gas. This is achieved by placing the lean oil cooler in a closed loop with a dedicated cooling tower. During the summer months, other columns could also benefit from such a dedicated cooling tower. These include the overheads of the Debutanizer, C_3/C_4 Splitter, and the Naphtha Splitter columns.

If several columns were to be cooled more efficiently in this manner, one or more temporary cooling towers could be used to provide a common supplemental cooling tower circuit for all the towers in the area (including the main fractionator).

It should be noted that several major process units face similar summertime cooling capacity limitations related to downstream separation columns. These include Crude, Vacuum, FCCU, Hydrotreater, Hydrocracker, Delayed Coker, and H_2SO_4 / HF Alkylation units.

There are numerous examples of similar summertime cooling limitations in the separation trains of many major petrochemical and complex chemical plants that were ameliorated by the supplemental cooling techniques described here.

4.3 Alkylation Unit Deisobutanizer

Distillation columns with total condensers are prone to one of the most undesirable effects of increasing summertime CW temperature. Most often, the column pressure control strategy uses a partially flooded condenser to ensure that the distillate remains at the bubble point so that there is no uncondensed vapor distillate. When condenser pressure rises in response to an increase in CW temperature, the relative volatility is diminished and fractionation suffers. The operator must then increase reboiler duty to restore bottoms purity, and the overhead reflux flow also must increase to maintain distillate purity. If column vapor load is already at the maximum, the feed rate must be lowered. A secondary effect is that the temperature of the vaporizing fluid in the reboiler increases as pressure rises. In some cases, if the heating medium temperature is insufficient, the reboiler duty can suffer to an unacceptable extent.

Such problems are often seen in alkylation unit deisobutanizers. Of course, they can occur in distillation columns throughout the refinery, e.g., in the debutanizer, C3/C4 splitter, or the naphtha splitter columns in gas plants that are attached to crude, FCC, hydrocracker, hydrotreater, delayed coker, and other units.

Shown below is an example where the tower pressure rises dramatically with CW supply temperature:

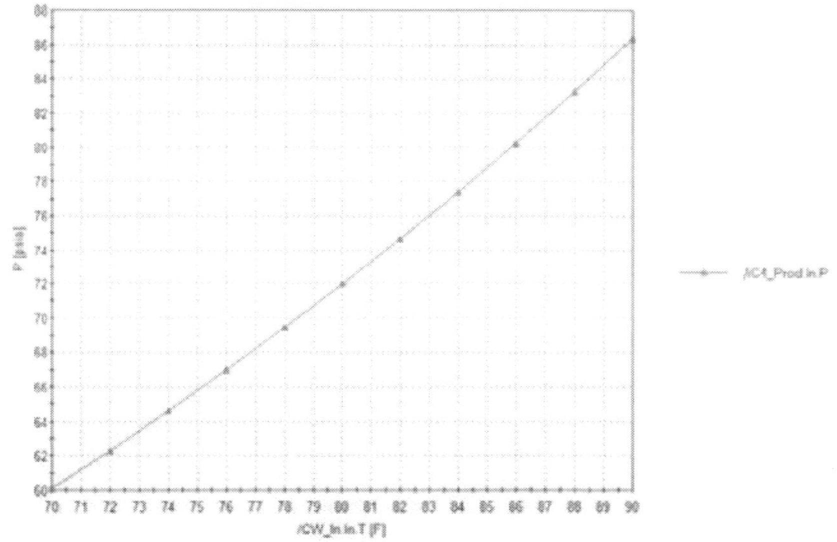

FIGURE 4: EFFECT OF COOLING WATER TEMPERATURE ON DIB TOWER PRESSURE

It can readily be seen that, in this example, reducing CW temperature by just 6°F reduced deisobutanizer tower pressure by as much as 8 psi, or 10%. In several past projects where alkylation unit capacity had been held back owing to this limitation, relaxing the CW temperature constraint by supplemental cooling enabled maximizing throughput during the peak summertime demand season when product margins were at their highest level.

5.0 PAST PROJECTS

We list below typical examples of column overhead cooling projects implemented by Aggreko Process Services. These were selected randomly from many dozens of similar highly successful project examples that have been executed over the past 15 years.

TABLE 1: EXAMPLES OF COLUMN OVERHEAD COOLING PROJECTS

AGGREKO PROCESS SERVICES - COLUMN OVERHEAD COOLING PROJECT EXAMPLES				
CLIENT	**UNIT**	**YEAR**	**WEEKS TO IMPLEMENT**	**BENEFIT/COST RATIO**
Gulf Coast Refiner	Crude 1 VDU Overhead	2006 & 2007	8	8:1
Gulf Coast Refiner	Crude 2 VDU Overhead	2006 & 2007	8	8:1
Gulf Coast Refiner	DEA Absorber Lean Amine	2014	4	20:1
Gulf Coast Refiner	MDEA Absorber Lean Amine	2015	4	30:1
Northern Refinery	FCC Main Frac Overhead	2004	6	6:1
Canadian Refinery	Deethanizer Lean Oil	2005 & 2006	8	8:1
Canadian Refinery	FCC Debutanizer Overhead	2005 & 2006	8	8:1
Canadian Refinery	NHT Debutanizer Overhead	2005 & 2006	8	7:1
Canadian Refinery	HF Alky Iso Stripper OH	2004 & 2005	6	5:1
Canadian Refinery	LERU Depropanizer Overhead	2003 to Present	6	5:1
Central U.S. Refinery	VDU Vacuum Condensers	2013	5	>50:1

It is worth mentioning that even in very cold climates, such as in Canada, the summertime swing in CW temperature can exceed quite significantly that in the Gulf Coast. This exacerbates the problems with tower pressure control and creates huge incentives for supplemental cooling projects of the type described here.

Another important issue relates to the practice that some operators follow when they use the diurnal temperature variation in CW temperature to try to maximize feed rates at night and then pull back during the day. This practice can be followed only when the tower pressure control functions flawlessly and the operator knows how to vary the set-points in anticipation of the forthcoming changes in CW temperature. In addition, changing the feed and reflux rates and reboiler duties invariably sets off composition transients that last for many hours. In the absence of model-predictive controls for the tower, these swings in product compositions can be hard to manage and, even with such advanced controls, we have found that tower feed rates hardly ever match what would be achievable with a steady CW temperature provided by a properly-designed supplemental cooling solution.

Column overhead cooling projects are executed very quickly when readily available, high-quality industrial-grade equipment is deployed, and up-to-date process engineering and project management support is provided in ways that supplement and complement the client's in-house capabilities. The elapsed time from the initial phone call to the commissioning is shown in column 4 of Table 1. All projects have very high benefit-to-cost ratios, far exceeding the normal threshold for internal capital projects. As noted earlier, it is generally difficult to obtain capital allocations for such temporary or seasonal projects. However, when there is a significant benefit/cost ratio, ranging from 5:1 to 30:1 for temporary projects of this type, our clients have found it far easier to obtain approvals using operating or maintenance budgets.

For each project, a Process Design Package is developed and furnished to the client. All process engineering details (PFDs and P&IDs, process simulation results, heat exchanger ratings, and hydraulic calculations) are provided. In addition, electrical details, equipment operational instructions, and operator training documentation are included. During commissioning, our technicians are available round the clock for ensuring proper operation. Following project completion, an after-action review is performed to document project performance and to record the timeline for all major events.

It is important to note that executing such projects requires a large inventory of industrial quality process equipment such as heat exchangers, refrigerated chillers, cooling towers, pumps, hoses, power generators, electrical cables and distribution panels. Aggreko tests all such equipment prior to deployment on each project. Heat exchangers are cleaned thoroughly and hydro-tested after each project using third parties. Refrigerated chillers and other rotating machinery also undergo periodic load testing to ensure performance and mechanical integrity. Field installation is performed by Aggreko's certified technicians who are highly experienced in executing such projects in a way that is consistent with client safety and environmental rules and requirements.

REFERENCES

1. Kister, H: "Distillation Operation", (McGraw-Hill, 1990)
2. Mathur, U and R.J. Conroy: "Successful multivariable control without plant tests", (Hydrocarbon Processing, June 2003)
3. Friedman, Y.Z.: "Control of Crude Fractionator Product Qualities During Feedstock Changes by Use of a Simplified Heat Balance" (Paper presented at American Control Conference, Boston, MA, 1985)
4. Mathur, U.: "Process Vessel Cooldown Apparatus and Method", U.S. Patent US9134064B2 (2015)

Troubleshooting a Thermosiphon Reboiler – Why New is Not Always Better than Old

Tuomas Ouni, Borealis Polymers Oy, Kulloo, Finland

Karl-Martin Hansson, Svante Welin, and Claes Hedvall
Borealis AB, Stenungsund, Sweden

Sarah Radovcich and Henry Z. Kister
Fluor, Aliso Viejo, CA

Presented at the Kister Distillation Symposium, AIChE Spring Meeting, New Orleans, Louisiana, March 31 - April 4, 2019

Summary

During an ethylene plant turn-around, a 38-year-old tube bundle of a vertical thermosiphon reboiler was replaced with a new, seemingly identical one. However, when the plant returned to full operation after the turn-around, the reboiler was not able to reach the same duty as the old tube bundle. The heat duty loss was substantial.

Systematic troubleshooting using detailed column and thermosiphon simulation, neutron backscatter, plant data reconciliation and flow regime analysis ruled out possible reasons such as thermosiphon inlet line plugging, or lowered heat transfer due to steam superheat, or the presence of non-condensables in the steam, leaving reduced heat transfer on the tube side as the strongest suspect.

To avoid plant shut-down, the issue was temporarily solved by switching from low pressure (1.8 barg) to medium pressure (7.2 barg) steam as heating medium. With the higher steam pressure and increased ΔT across the exchanger the original duty was successfully recovered.

An extensive heat transfer analysis using the HTRI software showed that the low heat transfer coefficients of the new tubes were due to film boiling. A good match to plant-measured heat transfer coefficients was obtained with a low fouling factor, about half the design. The analysis also closely predicted the successful results obtained with the 7.2 barg steam.

The heat transfer analysis, as well as historical record of finding clean tubes at the turnarounds, denied a theory that the old tubes performed better due to fouling that would cause them to transit from film to nucleate boiling.

To investigate the reduced heat transfer, light, laser and electron microscopy studies revealed that the surfaces of old and new reboiler tubes were very different in roughness and cleanliness, causing differing heat transfer characteristics. The gradual degradation of the tube surface, which occurred over the years, produced a much rougher surface than that of the new tubes. This rougher surface is likely to have provided a highly extended microsurface area and large number of nucleation sites that have increased the maximum heat transfer duty of the old reboiler. This theory is supported by a report from the old tube bundle's first year of operation when similar capacity problems occurred.

Reboiler Operation

Circulating vertical thermosiphon reboilers

Circulating vertical thermosiphon reboilers are compact, achieve high heat transfer rates, have relatively low fouling tendency, require little plot space and simple piping, and have relatively low capital cost and no pumping cost. This reboiler is usually the preferred type in the chemical and petrochemical industry[1]. Liquid to the reboiler is supplied from the tower base. Often, a preferential baffle divides the tower base into a constant-head reboiler draw sump and a level-controlled bottom draw sump (Fig. 1a). The baffle, which may be horizontal (Figure 1a) or vertical, preferentially diverts the colder liquid from the bottom tray into the reboiler, supplementing it with reboiler return liquid. Liquid overflowing the baffle into the bottoms sump is mostly the hotter reboiler return liquid.

Fig. 1b is a schematic of a reboiler tube. Liquid from the column base reaches the reboiler slightly subcooled due to the liquid static head. Upon entry, the liquid is preheated by sensible heat transfer only, with a relatively low heat transfer coefficient. Once the boiling point is reached, boiling starts, generating a two-phase mix of lower density than the liquid. Thermosiphon reboilers utilize the density difference between the liquid in the tower base and the mixed-phase fluid in the reboiler and return line to drive reboiler process flow. Unvaporized liquid returns to the tower base.

The lowest boiling zone is "bubble flow". The bubbles form by nucleation at the heating surface and are dispersed in the liquid. With more heating, more bubbles form, coalesce to larger bubbles, and then to slugs of vapor, termed "slug flow". Each reboiler tube slugs at a different pace so overall no instability develops. Upon further heating, an annular flow zone forms, where the liquid flows as an annulus and the vapor flows inside as a core. The vapor shear drags the liquid film up with it. Heat is removed from the wall by convection, with vaporization taking place at the vapor-liquid interface. Further heating boils off the liquid film, the wall dries up, and a mist flow zone sets in. Heat is transferred via a vapor film that has a high thermal resistance, and the tube wall temperature approaches the heating medium temperature. This condition is often termed "burnout", and may lead to metal overheating, fouling and poor heat transfer, so it needs to be avoided.

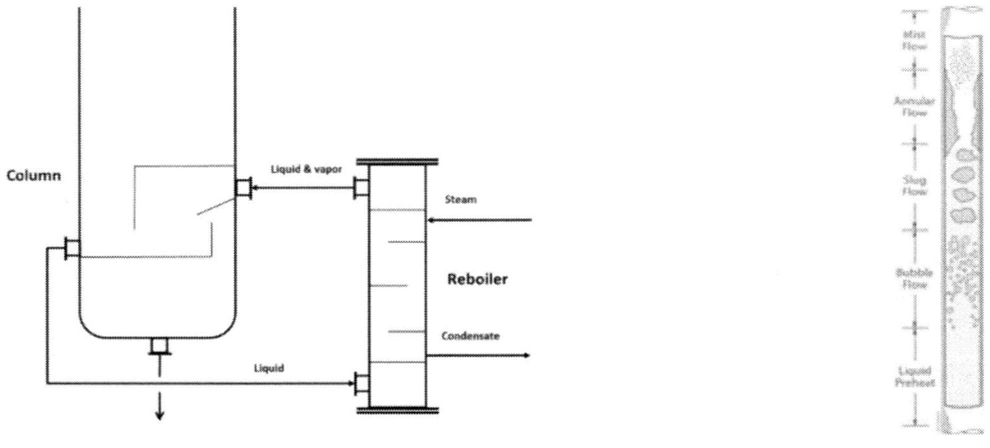

Figure 1. a) Circulating vertical thermosiphon reboiler b) Schematic of reboiler tube

At low liquid level in the reboiler sump, there is little driving head to push the liquid into the reboiler and the circulation rate is low. The little liquid that gets into the reboiler is totally vaporized, generating a dry-wall mist flow zone. Raising the liquid level intensifies circulation, lowers the fractional vaporization, lengthens the wet-wall boiling zone, and shrinks the dry-wall mist flow zone. Once the mist flow is eliminated, the reboiler reaches its maximum heat transfer rate. To keep a comfortable margin from mist flow, the industry's practice for pressure services has been to maintain the reboiler sump level (the top of the overflow baffle in Fig. 1a) at the top of the reboiler tubes[1].

The wet-wall regimes described above may not be attained in many services[2,3]. Nucleate boiling depends on the microscopic properties of the wall surface rather than the macroscopic geometry[3]. Vapor bubbles form in the superheated boundary layer within micro cavities in the heated surface. The wall superheat is the temperature difference between the wall and the local liquid saturation temperature. Bubbles grow at the wall and detach. Convective boiling occurs simultaneously with nucleate boiling. Convective boiling tends to dominate in the annular flow zone, under vacuum, and in wide-boiling mixtures, while nucleate boiling tends to dominate in the bubble flow zone, at higher pressure, and with narrow-boiling mixtures. When the wall superheat exceeds a certain value, known as the Leidenfrost point, a stable vapor film forms between the hot tube surface and the bulk liquid and blankets the wall. This is termed "film boiling". The vapor film has high thermal resistance, which drops the heat transfer rate. Film boiling is promoted by large temperature differences and clean tube surfaces. A transition region exists between the wet-wall boiling (nucleate and/or convective) and the dry film boiling region. This region has intermittent wet wall and dry wall portions and can be highly unstable.

To prevent film boiling in hydrocarbon and organic thermosiphon reboilers, it has been recommended[4] to keep the heat flux below 38,000 W/m^2 . This criterion is somewhat conservative, and an updated equation by Palen[5] has been recommended[6] .

Most reboilers are designed for nucleate boiling as that regime typically achieves high heat transfer rates and therefore minimizes the needed heat transfer area. In the nucleate boiling regime, the control of the heat flux is robust as the heat transfer increases linearly with increasing ΔT between surface and fluid (Fig 2).

Recently, experiences were reported in which reboiler performance at high heat flux did not improve after cleaning[3,6] because cleaning promoted a change from nucleate to film boiling. It was also reported that reducing the reboiler temperature difference by throttling the steam supply to a reboiler can change the boiling mechanism from film boiling to nucleate boiling, thereby improving heat transfer[7]. Another experience was reported[8] of loss of heat transfer following retubing of a circulating horizontal thermosiphon reboiler due to a reduction of nucleation sites. Later sandblasting restored the nucleation sites and reinstated good heat transfer. This article follows in this vein, reporting a detailed analysis of a case history in which replacing an age-roughened tube bundle by a brand new smooth one led to a large reduction in heat transfer rate.

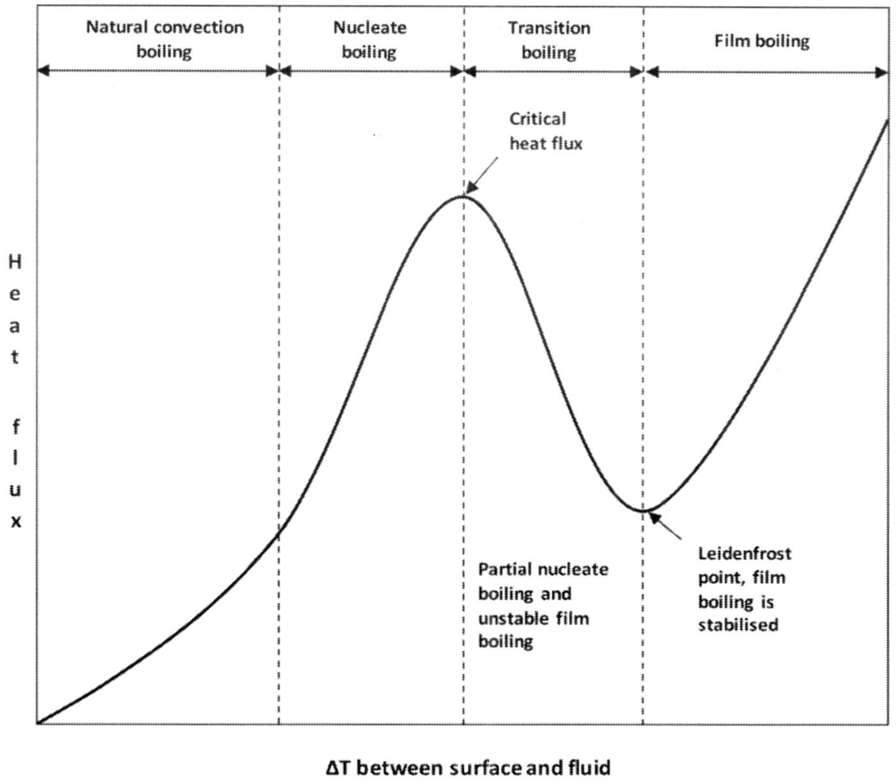

Figure 2. Boiling regimes as a function of the difference between surface temperature and liquid saturation temperature, based on Farber and Scorah[2], by permission courtesy of ASME

Reboiler Operating Problem

In this case history, a steam-heated circulating vertical thermosiphon reboiler is boiling a propane-propylene mix. Bottom tray liquid is preferentially diverted to the reboiler using a horizontal preferential baffle (Figure 1a). Steam flow to the reboiler is controlled via a control valve in the condensate outlet line, which keeps a condensate level in the shell. Increased condensate flow lowers the liquid level on the shell side, making more area available for condensing steam. Normal operating conditions before replacement in 2015 are shown in Table 1:

Table 1: Normal operating data for the reboiler before tube bundle replacement.

Shell side	Tube side
Steam condensing (at ~1.8 Bar(g))	Boiling C3 hydrocarbons
Tin 130°C	Tin 66°C
Tout 66°C (subcooled, note temperature approach)	Tout 66°C

During the cracker turnaround in 2015 the fixed tubesheet bundle was replaced due to several tube leaks earlier the same year as the turnaround, which caused plant shutdown for repair. The replaced bundle was 38 years old (it was installed in 1977). Prior to bundle replacement, the reboiler had excess capacity and was normally operating with high subcooling and condensate temperature approaching process side temperature. It was therefore decided to replace it with an identical one with only minor differences.

The old reboiler had 28 out of 725 tubes plugged and therefore the new one had slightly more area. The new reboiler had thicker tubes. Minimum wall thickness was increased from 2.11 to 2.77 mm.

Upon restart with the new tubes, the reboiler design capacity could not be achieved. As a maximum, about 11.5 t/h of steam could be condensed when 100% of the surface area was utilized (no condensate sub-cooling). Previously the reboiler could easily condense ~14.5 t/h of steam with high sub-cooling. Fig. 3 shows the temperature difference between condensate out and process side in before and after replacement.

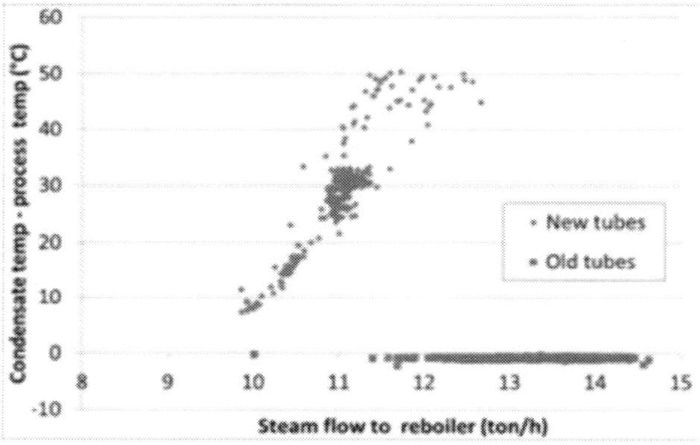

Figure 3: ΔT between condensate and process side before and after replacement of reboiler.

Troubleshooting

In order to solve the problem and recover reboiler capacity, systematic troubleshooting was conducted. The method adopted was to list all possible causes and then work through the list in a structured manner. Listing of all possible causes was done open minded by brainstorming.

Preliminary plant tests

First the most probable root causes were checked and excluded. Instrumentation was checked and calibrated. Process data was collected and mass and energy balance around the system were compiled

Steam side of exchanger was vented to atmosphere to remove possible light components.

Steam temperature to exchanger was adjusted to reduce steam superheat.

Feed rate was reduced and column bottom temperature increased to minimize light components on process side.

The reboiler was X-rayed to check for damaged baffles, assembly error, impingement plate assembly.

None of the above actions solved or explained the root cause of the problem.

Neutron Backscatter

Reboiler (heating side) scans showed the condensate level just above the draw nozzle, concurring with the little subcooling. In the tower, neutron scans were shot through both the reboiler draw sump and the bottom sump. One chord passed through the downcomer on the left of Figure 1a, the other to the right of the downcomer. Both showed the level in the tower bottom sump matching the level transmitter. Both reboiler draw sump scans showed frothy liquid of varying density. Near the top of the reboiler draw sump it showed mostly vapor, then the density increased to near the liquid density at the centerline of the 16" draw nozzle, then linearly declined to almost vapor density at the bottom of the sump.

The pipe taking liquid from the reboiler draw sump to the reboiler exited the tower horizontally for a short length, then dropped vertically 4 meters, then ran horizontally to the reboiler. Gamma scans of the reboiler draw pipe indicated that the pipe was full of liquid in the lower horizontal section, but only 75-85% liquid-full in the vertical section, indicating presence of vapor. The reboiler return pipe had about 30% vapor by mass, suggesting normal thermosiphon circulation.

Heat transfer coefficient from plant data

To calculate boiling heat transfer coefficients from plant data, it was necessary to subtract the reboiler area used for condensate subcooling. The subcooling area calculation was based on condensate level readings before and after reboiler tube replacement. The condensate level is measured by a local gauge. Following tube replacement, this gauge was closely monitored, so the subcooling area estimates are reliable. With the old tubes this gauge was infrequently monitored, so our pre-replacement subcooling area estimates are less accurate. Since the condensing duty far exceeds the subcooling duty, errors in the subcooling heat transfer have only a small impact on the boiling section heat transfer coefficients, and would only affect the pre-replacement coefficients.

Based on the collected plant data the heat transfer coefficients for the condensation area (area utilized above the condensate level) were calculated (Figure 4). Figure 4 shows conclusively that the heat transfer coefficients were largely reduced for the new tubes.

Fouling was considered as a root cause for the reduced heat transfer but review of inspection reports from previous turnarounds concluded that the exchanger had always been clean. Start-up data showed no entry of foulants during startup. During search for old data, a report from 1978, after the installation of the old reboiler, was found, indicating capacity limitation at less than 12 t/h steam to the reboiler. This is well in line with the capacity achieved for the new tubes (~11.5 t/h). Due to the lower plant throughput at that time it was then not considered as a major problem.

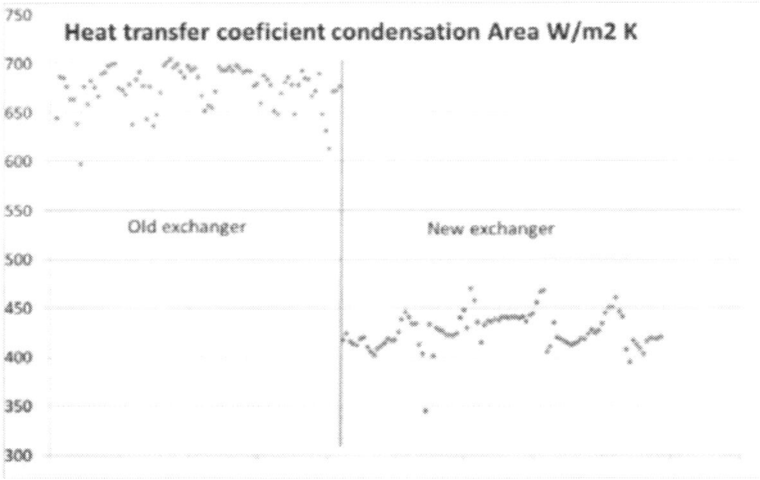

Figure 4 Heat transfer coefficient, condensation area for old and new reboiler developed over time.

Heat Transfer Analysis

Process simulation and reboiler circulation checks

An Aspen Plus simulation model of the tower and reboiler was built, using their EDR (Exchanger Design and Rating) tool, which allows detailed implementation of exchanger geometry. Inlet and outlet piping of the reboiler were included in the model. This allows estimating thermosiphon circulation rate, which is set by elevation between liquid levels and density differences between column and exchanger.

Distillation and reboiler simulation parameters were matched against plant data. Thermosiphon circulation rate was adjusted so that the pressure drop through the thermosiphon loop (including reboiler and nozzles) matches the hydrostatic pressure provided by liquid levels. According to the simulation model, reboiler heat transfer limitation lies mainly on the tube (process) side. The reboiler has too little surface area to perform its design duty with low pressure (1.8 barg) steam.

The calculations showed that the reboiler tubes operated in the bubbly, slug, and annular flow regimes (Figure 1), without approaching mist flow. The calculated tube outlet vapor content was 24% by weight, in good agreement with the 30% vapor fraction measured by the reboiler return line gamma scans. This fraction is consistent with common thermosiphon reboiler practices[1]. We conclude that the low observed heat transfer coefficients are not due to inadequate circulation.

Heat transfer: measured versus calculated

Detailed heat transfer calculations for the new tubes were performed using the HTRI reboiler rating program. The first step was to determine the fouling factor for the new tubes.

The original design fouling factors were 0.00035 m^2 °C/W and 0.000088 m^2 °C/W for the process and steam sides, respectively. To determine the actual fouling factor from plant data, Figure 3 was used. At 11 t/h steam the ΔT was 26°C, at 12 t/h approximately 47°C. We varied the fouling factor on the process side until we matched both. With the design (and expected) fouling being much less on the steam side, we applied a zero fouling factor on the steam side. A fouling factor of 0.00015 m^2 °C/W on the process side, about half the design value, gave a good match for these Fig. 3 data points. This is quite reasonable because the reboiler has been in service only a short time and only had time to lightly foul. The calculated heat transfer coefficients for the condensing area based on this fouling factor were around 450 W/m^2 °C, in excellent agreement with those determined from plant data in Fig. 4.

Nucleate or film boiling in new tubes?

To address this, it was necessary to construct a diagram similar to Figure 2 for the new tubes with the above-determined fouling factor. This was achieved by varying the reboiler steam pressure, which would vary the ΔT between the surface and the bulk temperature of the boiling liquid. The relevant surface temperature is the temperature on the inside of the fouling film. Fig. 5 is a plot of the average heat flux in the boiling/condensation zones against this ΔT. At the operating steam pressure of 1.8 barg, the ΔT between the surface and the process was 54°C. Fig. 5 shows that the operation was clearly in the film boiling regime.

Fig. 6 shows the average overall heat transfer coefficients for the condensation zone (no subcooling). It shows heat transfer coefficients of the order of 400-500 W/m2 °C, in excellent agreement with those determined from plant data in Fig. 4. Such low coefficients are consistent with film boiling.

A field test in which the steam pressure was reduced to 0.9 barg further supports the above conclusions. The heat transfer coefficient did not improve even though the delta T was about 12°C lower, in excellent agreement with Fig. 6.

Figs. 5 and 6 show that the Leidenfrost point, i.e., where the film boiling is stabilized for the new tubes occurs at a ΔT of 20°C between the surface and the bulk temperature of the boiling liquid. This coincides with a pressure of 0.8 bara in the reboiler. It may actually occur at a lower ΔT, but the instability in the transition zone made it impossible to determine it more closely.

Figure 5. Heat Flux vs. Surface to Process ΔT

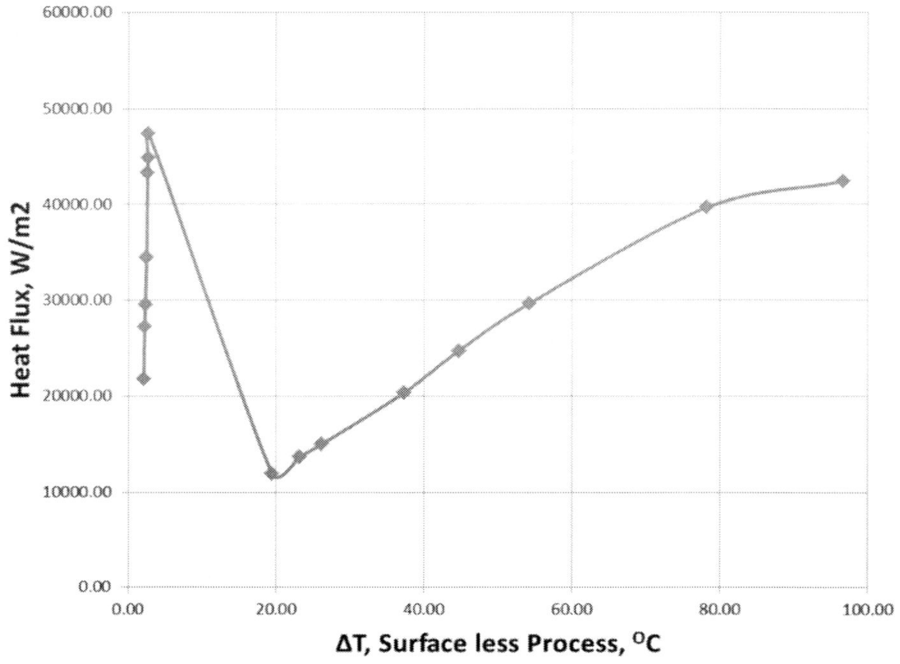

Figure 6 Overall Condensation Zone Heat Transfer Coefficients vs. Surface to Process ΔT

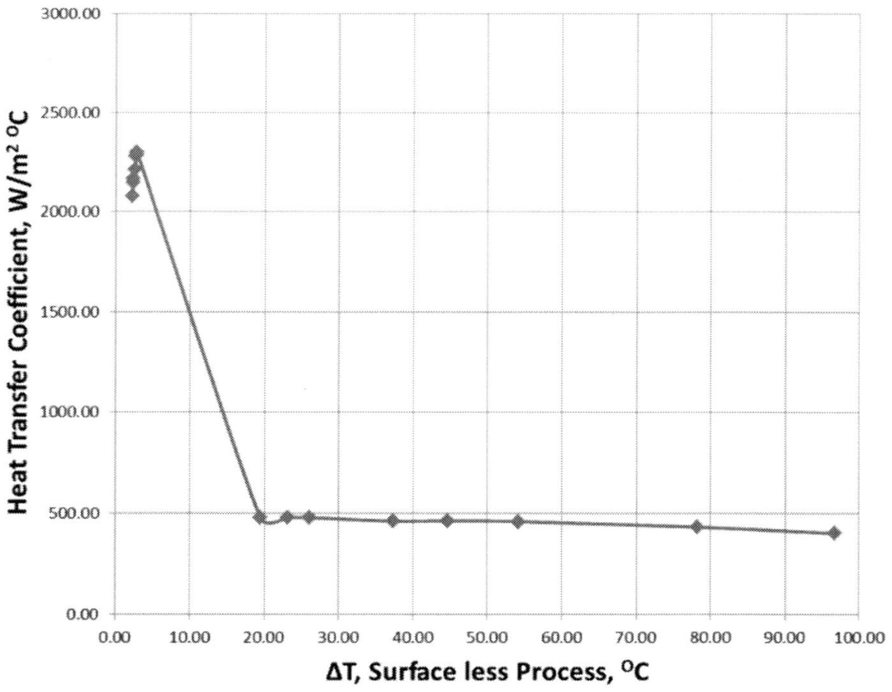

With 1.8 barg steam, a heat flux of 34000 W/m^2 ^oC was calculated directly above the condensate level. So the maximum heat flux over the reboiler was close, but not exceeding, critical heat flux of 38000 W/m^2 ^oC proposed by Kern[4]. Therefore, according to the rule of thumb, one would expect the reboiler to operate at the nucleate boiling region. Fig. 5 shows that caution is required when applying such rules of thumb, as they may only show part of the overall picture.

Did fouling cause the old tubes to perform better?

Experiences discussed earlier by Lestina, Serth, and Sofka[3,6] suggest that fouling can effect a transition from film to nucleate boiling, thus improving reboiler heat transfer. This may appear counter-intuitive, but the thicker fouling layer will lower the surface temperature and therefore the surface to process ΔT, moving the operating point to the left on Figs. 5 and 6. Fig.6 suggests that depending on the ΔT, the process side heat transfer coefficient increase may outweigh the higher heat transfer resistance due to the increased fouling. An early study by another engineering contractor attributed the better performance of the old tubes to their higher degree of fouling. Below we evaluate this conclusion.

Figs. 5 and 6 remain unaffected by fouling, as they are based on the difference between the surface temperatures on the inside of the tubes fouling film and the bulk temperatures in the tubes. For a fixed surface temperature, the steam condensing temperature will be higher with increased fouling (Fig. 7). In Fig. 7 the x-axis is the ΔT between the steam condensing temperature (not the surface temperature like in Figs. 5 and 6). Rdi are fouling factors, m^2 °C/W. Fig. 7 shows that as the tubes foul, the transition from nucleate to film boiling occurs at higher steam condensing temperatures (and pressures). With a high process side fouling factor of 0.0012 m^2 °C/W, about eightfold that calculated from plant data for the new tubes, the reboiler operation at a steam pressure of 1.8 barg would be completely within the nucleate boiling.

Our calculations confirmed that the higher fouling factors in Fig. 7 will improve the overall heat transfer coefficients. The overall heat transfer coefficients (including the subcooling sections) were 430, 540, and 610 W/m^2 ^oC for fouling factors of 0.00015, 0.00075, and 0.0012 m^2 °C/W, respectively. So far there is nothing to deny this theory.

The clincher came from Fig. 3, that shows the old tubes operating at steam rates as high as 14.5 t/h with condensate outlet temperatures very close to the process temperatures, a very high subcool. Our heat transfer calculations showed that for the high fouling factor of 0.00075 m^2 °C/W on the process side, the condensate temperature would be 90°C, about 25°C above the measured. Even for the unrealistically high fouling factor of 0.0012 m^2 °C/W on the process side, the condensate subcool temperature was still 9°C above measured. Both the 0.00075 and 0.0012 m^2 °C/W, especially the latter, are unrealistically high, as a review of inspection reports from past turnarounds concluded that the exchanger had always been clean. They were used here for the mere sake of testing the fouling theory.

These findings conclusively deny that fouling caused the good performance of the old tubes.

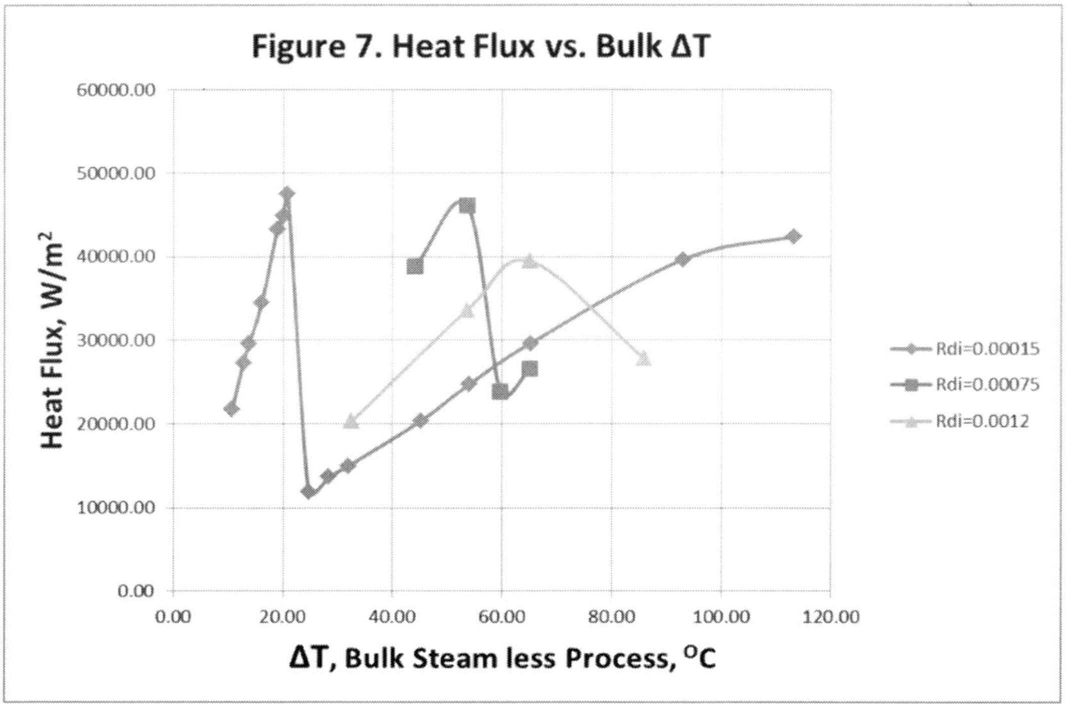

The Fix and Root Cause

As mentioned earlier, a plant test of reducing the inlet steam pressure to 0.9 barg showed that lower steam pressure had a negative impact on capacity. This is well in line with Figure 6, that shows little change in heat transfer coefficient. At the same time, a reduction in reboiler ΔT would reduce the heat duty.

Based on the plant test it was decided to construct a permanent line with medium pressure steam and perform a hot-tap to connect to the reboiler inlet piping. The reboiler capacity was successfully restored. The reboiler is now operating at full capacity with a steam pressure of 7.2 barg and subcooling of the condensate (condensate level 30 % of the tube bundle). ΔT between steam and process side increased from 65 ºC to 105 ºC but the heat transfer coefficient for the condensing area did not change with higher steam pressure and remained around 400-450 $W/m^{2o}C$ as can be inferred from Fig. 6.

Light, laser and electron microscopy

Samples from both old and new reboiler tubes were sent to a metallurgical laboratory for surface analysis. Fig. 8a and 8b show the difference in the roughness of the tube surfaces.

Confocal laser microscopy was performed for measuring the surface roughness of the tubes. Confocal laser scanning microscopy (CLSM) combines high-resolution optical imaging with depth selectivity, which allows to optical sectioning. With CLSM, visual sections of tiny structures can be viewed and 3-D structures can be constructed from the obtained images. In terms of absolute numbers, the roughness of the old tubes was 40-50 µm and the roughness of the new tubes 3 µm.

For chemical analysis of surface deposits, image scanning electron microscopy with energy dispersive X-ray spectroscopy (SEM/EDX) was performed. For the rougher old tubes, EDX-analysis revealed deposits or corrosion products containing significant amounts of Al, Si, Na and K likely present in an oxide form. For newer tubes with lower roughness, mainly carbon-deposits and Fe-oxide were detected. The roughened surface was therefore likely caused by the formation of deposits and corrosion products from Na, K, Si, Al impurities as well as from carbon-deposits/fouling.

Figure 8a (left) and 8b (right): Pictures showing the difference between surfaces of the new reboiler tubes (8a) and old reboiler tubes (8b)

Why was the heat transfer better with the old tubes?

Cai, Resetarits and Chambers[9] observations through viewing ports in a hydrocarbons kettle reboiler showed that at low heat fluxes the tube surfaces were not where nucleation began. Instead, nucleation started first on rusted segmental support baffles and rusted support rods. We believe a similar mechanism was active in our reboiler with the old tubes.

UOP markets HIGH FLUX™ heat exchanger tubing consisting of a porous surface bonded metallurgically to the tubes. UOP states "HIGH FLUX™ tubing operates exclusively in the nucleate boiling mode due to the porous surface. Saturated liquid is drawn into the layer by capillary action and vaporizes from the extremely large number of stable generation sites in the structure". UOP cites the highly extended microsurface area and large number of reentrant sites among the prime factors responsible for the tubes excellent performance.

While the surface of the HIGH FLUX™ tubing is far superior to the pitted surface of the deethanizer reboiler old tubes, Fig. 8b shows that the old tubes inadvertently share some of the performance-enhancing features. A calculation based on HIGH FLUX™ tubes showed that had the reboiler contained those, it would have readily achieved the high subcool in Fig. 3.

There are tens of thousands reboilers in newly-constructed plants, or fitted with new tubes, yet this poor heat transfer appears uncommon. This may be because most reboilers operate in the nucleate boiling regime, where the surface effects are far less critical.

Only one literature source[8] reported a similar experience, and in his case the higher steam pressure made the loss of reboiler capacity marginally worse. Fig. 6 shows that if the initial operation was in the transition zone, at less than 20°C ΔT between the surface and the bulk, then the reduction in heat transfer coefficient at the higher pressure would outweigh the gain from the higher steam to process ΔT. In this case too, the solution was roughening the tube surface.

Conclusions

This case shows that the surface roughness developed in an old exchanger over time had a significant beneficial effect on film boiling heat transfer. The old tubes had been mechanically roughened, due to corrosion and deposits during 38 years of operation. This roughening provided an extended microsurface area, porosity, and large number of nucleation sites, which under film boiling highly improve heat transfer compared to bare tubes. This resulted in a reboiler capacity increase by more than 50% compared to bare tubes.

Under film boiling, fouling has been proposed as a supporting mechanism for heat transfer improvement [3,6]. However, our calculations showed that even with much more severe fouling than occurred in our reboiler, the heat transfer improvement due to fouling would have been smaller than observed. Further, turnaround inspections always found the tubes clean, arguing against fouling being a significant contributor.

To the best of our knowledge, this is the first time that a full heat transfer analysis of the mechanism in which surface roughness improves heat transfer under film boiling conditions , supported by field data, is reported in the literature. The knowledge gained would help engineers troubleshooting and revamping reboilers operating in the film boiling regime. Boiling regime diagrams similar to Figs. 2, 5 and 7, as well as close reviews of subcooling data, like Fig. 3, were demonstrated to be invaluable analytical tools.

This case serves as a lesson for engineers changing old heat transfer equipment to new. The smooth and clean tubes in the new equipment may not always act positively on overall heat transfer. Checking the reboiler rating and ability to achieve its duty should be carefully performed also in cases of like-to-like replacement.

References

1. Kister, H. Z., *Distillation Operation*, McGraw-Hill, NY, 1990
2. Farber F.A. and Scorah R.L., , *Heat Transfer to Water Boiling Under Pressure*, Trans. ASME, 79, p. 369, 1948.
3. Lestina, T., Sofka, C., *The Boiling Point*, Hydrocarbon Engineering, September, p. 27, 2015
4. Kern, D. Q., *Process Heat Transfer*, McGraw-Hill, NY, 1950
5. Palen, J. W., *Shell and Tube Reboilers, in Heat Exchangers Design Handbook*, Vol. 3, Hemisphere, NY, 1988
6. Serth, R.W., Lestina, T. G., *Process Heat Transfer, Principles, Applications, and Rules of Thumb*, 2nd Ed., Academic Press, Waltham, MA, 2014
7. Hagan, M. D., and Kruglov, V. N., *Understand Heat Flux Limitations on Reboiler Design*, Chem. Eng. Progr., November, 24, 2010.
8. Lieberman, N. P., *Improving Heat Transfer in Reboilers and Condensers*, PTQ, Q1, 39, 2016
9. Cai, T., Resitarits, M., and Chambers, S., Kettle Reboiler Hydraulic Studies, DA2014, p.282, 2014
10. UOP, "Advanced Heat transfer Technology with HIGH FLUX™ Tubing", Bulletin EP4063B, Tonawanda, NY

545200 Rapid 1st-Pass Design of Dividing-Wall Columns Via Acyclic Simulation Method

The "acyclic" method was proposed by Navarro et al in 2012 to simulate thermally coupled distillation sequences such as side-strippers. This approach allows coupled columns to be simulated once-through without recycle streams, by emulating the pre-fractionator overheads as a refluxed section with superheated vapor, and the pre-fractionator bottoms as a reboiled section with sub-cooled bottoms. The superheated vapor and subcooled liquid, in turn, are replaced by saturated vapor and liquid streams and parallel heat streams transferring the pre-fractionator condenser and reboiler duties to the main column.

The acyclic methodology was implemented as a process simulation procedure that can be used to easily develop 1st-pass designs for dividing wall columns (DWC). The acyclic simulation runs much more easily than conventional coupled column simulations with recycle streams between the pre-fractionator and main column, facilitating sensitivity studies and preliminary optimization of the number of stages in each section of the DWC, as well as the liquid split ratios.

An optimized acyclic simulation is readily converted to and initialize the corresponding fully-coupled ("cyclic") column simulations with recycle streams between the pre-fractionator and main column. The reboiler duties vs. product purity calculated by the acyclic method agree within a few percent of the rigorous cyclic simulations. Final simulations using the cyclic simulations are recommended, to allow for final optimization, dynamic simulation to evaluate control strategies, and calculation of hydraulics which can then be turned over to DWC column internals vendors.

An example is shown of a DWC modeled using Aspen Plus first using the acyclic mode and afterwards converted to a rigorous fully-coupled 2-column model (cyclic mode), and the results are compared.

551763 Four High Purity Products from Dividing Wall Column

Dividing wall columns typically combine two columns into one, produce three cuts and are known to save 20-30% in OPEX and CAPEX.

After operating successful DWCs many refiners are approaching us to develop and install more complex DWC configurations. These configurations can produce four high purity cuts and can replace three columns.

The case study presented in this paper is from the Isomerization process involving Deisohexanizer column. Many Isom units are lined up to produce a C6 cut, which is used as a commercial solvent in this DIH Column. The production of C6 cut comes at an expense of a decrease in overall RON from the gasoline pool. As of now, usage of DWCs is limited to combining of two columns (DIH Column and C6 cut column) to produce three cuts (Light Isomerate, Heavy Isomerate and C6 Cut). In absence of a recycle stream of low RON components to the Isomerization reactor, there is a significant penalty of RON. The revamped DWC produces four cuts in additional to light and heavy isomerate products, the column produces C6 cut and a stream of low RON components as a recycle to the ISOM reactor. This four cut DWC fits in existing DIH column revamps. This innovative DWC configuration not only provides additional revenue of C6 product but also minimizes any impact of RON in the gasoline pool.

Thermodynamically, the four cut DWC combines the features of a middle and a top dividing wall column to achieve the required separation.

550304 Dual-Dividing Wall Column R&D and Deployment

Dual-Dividing Wall Column R&D and Deployment

Robert Piszczek, Michael L. Hergenrother, Sundar Narayanan, Rustom Billimoria

ExxonMobil Research and Engineering (EMRE) Company

EMRE discovered simplified design configurations for Multi-Partition Dividing Wall Columns (MP-DWC) that contain two dividing walls and are referred to as Dual-Dividing Wall Columns (DDWC). DDWC is the next generation of Dividing Wall Columns (DWC) technology that utilizes Process Intensification (PI) to produce up to six high purity products (>99 wt%) in one vessel at reduced CAPEX, OPEX, and plot space compared to multiple conventional columns and/or DWCs in series.

Multiple process configurations with DDWCs were developed and screened for a potential first commercial deployment for a new chemicals project. Numerous technical challenges were encountered, mitigated, and no showstoppers are anticipated.

Cost estimates indicate total erected cost (TEC) savings of 30% for a 6-Product DDWC compared to the next best alternative consisting of two DWCs and one conventional column in series. A risk mitigation strategy was developed to reduce the potential risk exposure for the first deployment of DDWC technology. As a result, a lower consequence 5-Product DDWC configuration with prorated CAPEX savings of 20% was selected as the lead option for further development, evaluation, and detailed design in parallel to the Stage 3 base process configuration for the project. Also, energy savings of up to 25% were projected but depend on each specific application and DDWC design philosophy.

Robert Piszczek, P.E. graduated from Rutgers University with B.S. and M.S. in chemical engineering. He has over 17 years of experience in process engineering in the areas of distillation (9 yrs), process control (6 yrs), gas treating and wet treating (2 yrs). He is currently on assignment as the Distillation Specialist for ExxonMobil Asia Pacific PTE LTD for Singapore.

551761 Development of Packing Design for Distillation in Rotating Packed Beds

Fluid separations performed in conventional columns are limited by the gravitational field in terms of applicable liquid and vapor loads. Difficult separation processes or high throughputs often result in large column dimensions and high investment cost. However, flexible and modular equipment is of growing importance for meeting challenges in the chemical industry. Rotating packed beds (RPBs) are a promising example of flexible and modular contacting equipment employing HiGee (high gravity) technology. The centrifugal force imposed by the rotation of an annular shaped packed bed located in a stationary casing superimposes the gravitational force and results in reduced film thickness and enhanced mass transfer. This allows the implementation of comparable separation performance in smaller equipment with a higher capacity compared to a conventional column [1].

So far, the range of available packings for RPBs is rather limited and mainly wire mesh and metal foams are employed for various separation processes. For these isotropic packings, the cross-sectional area and the gas capacity factor, commonly represented by the F-factor, is continuously changing along with the radius of the packing. As recent gamma-ray tomography showed, a significant decrease of liquid hold-up with increasing radial length of the packing and rotational speed [2]. These variations decrease the overall performance of the packing. To overcome these limitations a newly developed ZickZack packing, like the concept of a tray-like structure in rotating zigzag beds (RZB), was developed. Contrary to RZB, the ZickZack packing is a regular structure, which is placed within the rotor of an RPB.

The current study represents a comparison of the separation efficiency and the pressure drop of conventional isotropic packings and the ZickZack packing in an RPB. Separation of an alcohol-water mixture under total reflux distillation operation at atmospheric pressure is performed. The experimental results show that the ZickZack packing not only improves the separation efficiency in the packing but also improves the overall separation performance of the RPB without additional pressure loss within the investigated vapor and liquid loads.

References

[1] Neumann et. al., Chem. Eng. Res. Des., 2018, 134, p. 443 – 462.

[2] Groß et. al., Proceedings of 9th World Congress on Industrial Process Tomography, 2018, Bath, UK, p. 831 – 838.

The research is done in cooperation with the Institute of Sustainable Process Technology (ISPT) in the framework of the project ImPaCCt (Improved Process Performance by Process Intensification in Centrifugal Contactors).

551762 Bottom Dividing Wall Column in a BTX Unit

DWC's have been successfully applied in BTX separation units before, but almost all designs have been of middle DWC type i.e. where the dividing wall is placed in the middle part of the column. In a conventional post-fractionation section of a BTX extraction unit, a sequence of two distillation columns in series is used to get three high purity products – a B/T column, where benzene is recovered as top product with a purity level of 99.9 wt%, and a T/X column where toluene is recovered at the top with a purity level of 99.5 wt% min. C8 aromatics (mixed xylenes + EB) are drawn as the bottom product of the T/X column with a typical purity range of 95.0 – 98.5 wt%, depending on the quality of feed sent to the unit. Combining these two columns is particularly challenging due to high purity level of the products while still justifying the energy saving benefits associated with the DWC design.

GTC Technology has designed the first of its kind - Bottoms DWC for this application, and the presentation will touch upon the following items:

- key aspects of B-DWC design,
- its advantage over middle DWC for this application, and
- Control scheme needed to successfully operate this column.

The above mentioned column is currently being fabricated and is expected to be operational in Q4 of 2019.

545149 Purification of Styrene from a Styrene/Mixed Xylene Mixture By a New Technique Combining Distillation and Crystallization

Stripping crystallization (SC) is a new separation technique which combines distillation and crystallization. Basically, SC is operated at a triple-point condition, in which the liquid is simultaneously vaporized and crystallized due to the three-phase equilibrium. In essence, the process is continued until the liquid phase is completely eliminated and only the pure solid crystals remain in the feed. Thus, crystal washing is not required since only pure solid crystals remain in the feed and no impurities are adhered on the crystal surfaces at the end of SC. In the previous research, SC has been successfully applied to separate p-xylene from the mixed xylenes by Shiau et al. (2005, 2008, 2013).

SC is introduced in this work to purify styrene (ST) from the multi-component system consisting of ST, ethylbenzene (EB), p-xylene (PX), m-xylene (MX) and o-xylene (OX). Styrene is one of the most important aromatic compounds and is used extensively in the manufacture of polystyrene. Due to the close boiling points of ST, EB, PX, MX and OX, it is rather complicated and energy-intensive to separate them by conventional distillation. A thermodynamic model is developed to determine the three-phase equilibrium conditions and direct the SC experiments. The experiments show that SC can be effectively applied to purify ST from the mixture. However, as EB is more easily incorporated into the ST crystal lattice compared to PX, MX and OX, it becomes more difficult to further purify ST for a higher concentration of EB in the mixture.

References:

1. Shiau, L. D. "Purification of styrene from a styrene/ethylbenzene mixture by stripping crystallization", *Ind. Eng. Chem. Res*, **57**, p6759-6765, 2018.

2. Shiau, L. D., Wen, C. C. and Lin, B. S. "Separation of p-xylene from the multicomponent xylene system by stripping crystallization", *AIChE Journal*, **54(1)**, p337-342, 2008.

3. Shiau, L. D.; Liu, K. F. "Investigations into the effects of the cooling rate on stripping crystallization", *Ind. Eng. Chem. Res*, **52**, p1716-1722, 2013.

Keywords: crystallization; distillation; purification; styrene

551764 How Geographical Location Changes the Configuration of a Dividing Wall Column?

Distillation as such is an energy intensive unit operation. Dividing wall columns have gained credibility in refineries and petrochemical complexes as an attractive alternative to conventional distillation sequences. The dividing wall helps to prevent remixing of components thereby improving the efficiency of the multi-component separation and lowering the energy requirements. DWCs have been proven to provide energy and capital cost savings as compared to conventional multi-component distillation.

Plant geographical location plays a key role in deciding the DWC configuration since utility costs and CAPEX vary by geography. The article looks at the various configurations and establishes a baseline for choosing a DWC configuration based on the client's utility arrangements and CAPEX requirements. The configurations that has been studied for this purpose include – a Middle DWC, a Top DWC and a Bottom DWC. An example of BTX separation is presented for each configuration to discuss the different technical solutions required in different parts of the world.

Middle DWCs: These are the most common and energy efficient configuration. However, the column typically operates under the highest level of utility for reboiling duty and the lowest level of utility for condensing duty. For example, the heating duty shifts from low pressure (LP) steam to medium pressure (MP) steam for a BT separation. For geographical locations where there is a significant difference in the price for LP and MP steam, a Middle DWC will not produce the necessary cost savings.

Top DWCs: These columns typically provide lower overall utility savings as compared to a middle DWC. The column will however have two separate condensing temperatures, each requiring a separate condenser. The middle product overhead vapors can be used for heat integration purposes, which offsets the higher equipment costs.

Bottom DWCs: Similar in design to a Top DWC, the 'dividing wall' is located at the bottom. These columns hence accommodate the use of two separate reboiling utilities. This configuration provides significant energy savings in geographical regions with high utility costs.

AUTHOR INDEX

AUTHOR INDEX